THE WEST
AFRICAN HERBARIA OF
ISERT AND THONNING

PETER THONNING (1775–1848)

Painted in September 1841 by E. Bærentzem. Portrait in the possession of
Mrs Birte Jacobson of Jylland, a descendant of Thonning.

PLATE 1

THE WEST AFRICAN HERBARIA OF ISERT AND THONNING

A taxonomic revision and
an index to the IDC microfiche

F. N. HEPPER

The Herbarium, Royal Botanic Gardens, Kew

BENTHAM-MOXON TRUST
ROYAL BOTANIC GARDENS
KEW

Published by the Bentham-Moxon Trust, in association with the Carlsberg Foundation. Obtainable from the Bentham-Moxon Trust, Royal Botanic Gardens, Kew, Richmond, England price £6.50 plus postage

Printed in Great Britain by
Robert MacLehose and Company Limited
Printers to the University of Glasgow

Dedicated

to the memory of all plant collectors who died
in West Africa, especially

P. E. Isert

J. von Rohr

O. H. Smith

L. Mortensen

CONTENTS

7

ILLUSTRATIONS

INTRODUCTION

Between 1783 and 1803, at a time when little was known of the tropical African flora, two Danes, P. E. Isert and P. Thonning, made very important collections of plants in West Africa. About 2,000 specimens, representing over 600 species were received by the Professor of Botany at Copenhagen, Martin Vahl, who distributed many of them to botanists of the day in other countries. Thus it happens that several hundred sheets are to be found in the herbaria of ten other countries besides Denmark, where the major part of the collections is housed at the Botanical Museum, Copenhagen. As many, indeed most, of the plants were described as species new to science, these collections are rich in type specimens of all categories. While working on the revision of the 'Flora of West Tropical Africa', I realised that the identity of all the specimens ought to be checked and the nomenclatural synonymy worked out; this has now been accomplished. Moreover, it was possible to trace many of the duplicate sheets at Copenhagen, Paris, Berlin and Geneva which have been photographed for microfiche sets that are now readily available for consultation in major herbaria, as well as other specimens at Leningrad, Munich, Florence, the British Museum and Kew. Hence the following account is both a taxonomic revision and an index to the location of the specimens and the relevant microfiche.

During a visit to Copenhagen in 1966 I was able to go through the whole collection. Subsequently I worked directly from specimens on loan at Kew or from the microfiche, double-checking the identifications from the specimens if there was an element of doubt in critical species. I have also visited Stockholm, Florence, Geneva, Munich and Paris in my quest for duplicates.

From the historical aspect it is interesting to note how the duplicates came to be lodged in their present herbaria. A full history would be a study in itself, but a few notes will suffice for our purposes.

The Berlin specimens are duplicates of Isert's collection which were sent to Willdenow from Copenhagen in 1799, that is before Thonning's material arrived in Europe. However, Isert's specimens at Munich went there in 1813 with the herbarium of Schreber who presumably obtained them directly from Vahl before the former's death in 1810. At Stockholm, the Isert sheets appear to have been donated by Willendow, while Thonning himself seems to have presented his own material. Those in Leningrad may have been distributed at a later date. Vahl sent Thonning's duplicates to Paris where many of them are still available in Herb. Jussieu. However, others were in Desfontaines' collection which was bought by Webb who bequeathed his herbarium to the Grand Duke of Tuscany for the University of Florence. Many other specimens found their way to De Candolle at Geneva and the history of one sheet is especially interesting. De Candolle obtained a specimen of *Polycarpaea stellata* from Puerari, who died in that city, and later Jacques Gay was presented by De Candolle with a fragment which came to Kew when Gay's collection was purchased in 1868. Incidentally, this is one of two specimens so far traced at Kew. The other specimen is *Thonningia sanguinea* which was sent to Kew from Copenhagen by Prof. F. M. Liebmann in the

early 1850's for Hooker's revision of Balanophoraceae. A duplicate is also at
the British Museum (Natural History) and this is the only one traced there,
presumably because the Napoleonic wars in the early years of the 19th
century precluded botanical exchange from other European countries.

As mentioned above, the principal collectors were Isert and Thonning with
a few specimens gathered by von Rohr and Mortensen, but many sheets
have been attributed to others who are known not to have collected in West
Africa. Thus the names of Vahl, Hornemann, L'Héritier, Pflug, Ryan,
Banks, Hofman Bang, Puerari and de Jussieu appear on them, often with the
locality recorded as "e Guinea". They were presumably distributed by those
named but collected by Thonning, in which case they would be duplicates
of the numbered specimens. However, in this work I have normally entered
them separately under "e Guinea" without indicating the presumed
collector or their possible type status.

There are 474 names based on type specimens in the Isert and Thonning
herbaria, and 203 of the epithets are currently in use either in the original
genus or by transference to another. This is out of a total of 610 species listed
in the following paper, which indicates the richness in type specimens and the
importance of these collections to taxonomic botany.

The totals of herbarium sheets in the various herbaria are as follows
(see Appendix V a–i for details):

Berlin-Dahlem (B): 71 sheets, all Isert's in Willdenow's Herbarium.
British Museum (BM): 2 sheets, Thonning's.
Copenhagen (C): 1484 sheets, comprising P. E. Isert 283 sheets,
P. Thonning 1027 sheets, L. Mortensen 23 sheets, and 151 sheets
attributed to others, although presumably most of them were collected by
Thonning and some by Isert. A few sheets collected by C. Smith (Congo),
J. G. König and J. P. Rottler (India), and H. Smeathman (Sierra Leone)
are included for the reasons stated. (These figures compare with those
given by Junghans (1961, 1962): total number 1467 sheets, comprising
Isert 279, Thonning 990, Mortensen 24, von Rohr 7, C. Smith 2,
L'Héritier 4, Pflug 1, Ryan 1, Anon. 135, König 7, Rottler 1).
Florence (FI): 66 sheets, all Thonning's although normally marked 'Vahl
1804'.
Geneva (G–DC): 30 sheets, mostly Thonning's.
Kew (K): 2 sheets, Thonning's.
Leningrad (LE): 8 sheets, Thonning's.
Munich (M)*: 3 sheets, Isert's.
Paris (P–JU): 172 sheets, all Thonning's although usually marked Vahl.
Stockholm (S): 55 sheets, mainly Thonning's with a few Isert's.
Vienna (W): none traced, but several specimens were grown there from
Isert's and Thonning's seed.

Other specimens are sure to be traced in any or all of the above herbaria,
especially Munich and Geneva, where, owing to lack of time, I was not able
to delve as extensively as in some others. Taxonomists are requested to
notify me of the whereabouts of additional specimens and also of type
designations.

* Munich is latinised Monachum and should not be confused with Monaco, as did
Junghans (1961: 319).

Biographical Notes

Paul Erdmann Isert (born Angermünde, Brandenburg, of German parents, 20 October 1756—died of fever West Africa, 21 January 1789) went in 1783 to Danish Guinea, the coastal parts of present-day Ghana and Dahomey, where he was a physician. He left on 7 October 1786 via the West Indies, but returned in July 1788 to start plantations in order to combat the slave trade, which grieved him greatly; alas he died in January 1789.

Peter Thonning (born Copenhagen, 9 October 1775—died Konferent-straad, 29 January 1848) studied medicine and in 1799 he went with O. H. Smith to the part of Danish Guinea that is now in Ghana where, until his return to Denmark in 1803, he tried to fulfil Isert's plan by founding plantations of tropical crops, and studying natural resources and dye plants. He collected numerous wild plants with detailed notes and descriptions, but unfortunately his own set of the material was destroyed during the bombardment of Copenhagen in 1807 and Thonning lost heart in botany and became a customs officer.

Soon after Thonning's return Vahl published some new species, but it remained for H. C. F. Schumacher, Professor of Anatomy at Copenhagen, to write up the material Thonning had previously given to Vahl and publish it as a book in 1827. The same work was apparently re-published in 1828 and 1829 in two parts of a Danish scientific journal. Unfortunately this has given rise to much discussion as to whether the book was a pre-print or whether the date of 1827 should read 1829. Botanists at Kew and Copenhagen still take the view that 1827 is the correct date, until it can be proved otherwise, although Christensen stated that "it is highly improbable that the re-print was finished in 1827. Further it may be noted that the earliest review of the paper seen dates from late in 1829" (Flora Malesiana ser. 1, 4(5): CCXII (1954)). However, both sets of page numbers are quoted in this paper, although Junghans cited only the pagination of the pre-print.

Another point of dispute is the authorship of the plant names. Thonning named some species and prepared long and accurate descriptions of many others, but they were largely published by Schumacher. The latter also supplied names and descriptions when necessary and often added further notes. Schumacher carefully attributed authorship to each paragraph by concluding with "Th." or "S". Thus the Kew tradition, which was followed by Junghans, has been for a name which bears a diagnosis initial "S", but with a description concluded with "Th.", to bear the authority "Schum. et Thonn.", except when the epithet *thonningii* is used: the authority "Schumacher" is then used alone. Similarly, a single authority is attributed when the diagnosis and the description are both initialled "S" or "Th.".

Chronological lists of the publication of new species from both Isert's and Thonning's collection are given by Junghans 1961 : 314–318.

A few specimens collected by other Danes are also included in this account as they have been regarded in Copenhagen as part of this collection.

Julius Philip Benjamin von Rohr (born c. 1737, died 1793 on the way to West Africa) collected a few specimens in West Africa. His principal collecting area was the West Indies and South America. According to

Mrs Fox Maule (1974: 188) von Rohr's 7 specimens were the first to be received by Vahl from West Africa.

Ole Haasland Smith (died Akwapim, West Africa, February 1802). He sailed to Danish Guinea with Thonning in 1799 and assisted him in his investigations as "friend and travelling companion" (see p. 63). Perhaps he collected many of the plants attributed to Thonning and it is sad that his only memorial is the genus *Hoslundia* (sic).

Lars Mortensen (born 1798, died West Africa 1821) collected some 20 specimens in West Africa while he was a regimental surgeon. Schumacher evidently received this material which was added to that of earlier collectors

FORMAT

In Junghans' useful enumeration of the Isert and Thonning collections at Copenhagen published in 1961 and 1962 (q.v.), he based his list primarily on Willdenow, 'Species Plantarum' and Schumacher, 'Beskrivelse af Planter', using the same nomenclature. These names were published by the following authors: A. and A. P. De Candolle, Dunal, Geiseler, Horneman, Isert, Jacquin, Persoon, Poiret, Schumacher, Sprengel, Thonning, Vahl and Willdenow. However, since these names were published in 1827 or before, few of the species are now recognized by the same names.

Over a period of some years, I have, therefore revised the whole collection taxonomically and nomenclaturaly and published in Kew Bulletin from time to time relevant notes and new combinations. In the present paper the species are arranged alphabetically by family and genus under their currently recognized botanical name, with relevant synonyms, but excluding *nomina nuda*. Junghans dealt with the material at Copenhagen only, while this paper accounts for all the duplicates that have been traced in ten other herbaria, as well as indexing the microfiche when available. Vernacular names and translations of field notes, economic notes and place-names are also given here. Some facsimiles of handwriting and labels are given, but reference should be made to further examples in Junghans (1961: Figs. 1–6) since he dealt with the old manuscripts preserved at Copenhagen.

Specific names

In order to be concise, the basic reference to the species is the recently completed Revised Edition of the 'Flora of West Tropical Africa' (FWTA.), which should be consulted for synonymy, with later amendments when necessary. The older 'Flora of Tropical Africa' (FTA.) is only cited if it provides useful synonymy actually based on these West African collections. For the sake of complementarity many of the works cited have been abbreviated in the same manner as that adopted by Junghans and a bibliography is included for those works not given in full. Altogether 610 species are accounted for, compared with 565 in Junghans' papers.

Exsiccatae

To facilitate citation of these specimens in future taxonomic works, such information as is known is set out for the first time in the customary manner with locality, date, collector, and number. Then follows the herbarium location using standard abbreviations, with the number of sheets extant, whether a type specimen, and microfiche reference if any. Note that the citations of exsiccatae are carefully placed under the relevant name or synonym, although in some cases the specimens must be duplicates of the same gathering.

About 140 sheets which were previously undetermined have also been identified (after nearly 200 years!) and added to the citations. Many of them were seen by Junghans, but as a non-botanical bibliographer he was not in a position to identify them.

Localities

The original spelling of the localities is given in the citation and the modern equivalents may be found by reference to the gazetteer on blue paper in Appendix II (p. 171) and Maps 1 & 2. All of Thonning's localities are in Ghana only, while Isert is known to have collected in Ghana and Dahomey. No Togo locality is given. The maps (p. 174) show the general region and the location of the smaller places in present-day Ghana. When no precise locality is known "Guinea" is entered in the text, although these specimens are certainly from Ghana, except possibly for some of Isert's which could be from Dahomey.

Dates

The precise year of collection is known for only some of Isert's specimens, who wrote it on the sheets. Isert collected between 1783 and 1786, Thonning between 1799 and 1803, von Rohr before 1793 and Mortensen before 1821.

Numbering

I have applied the number appearing in Schumacher's MS to Thonning's herbarium sheets at Copenhagen. Many of the sheets in fact already bear these numbers and it will be convenient in future if authors constantly cite them. The same numbers may also apply to the specimens which appear without Thonning's name and entered here as "e Guinea". This also applies to the distributed duplicates which are seldom written up with Thonning's name. Isert's sheets were not numbered, except for a few added later which are of no significance to us.

Types

The current terminology laid down in the Code of Botanical Nomenclature provides for many categories of type specimens; holotypes, syntypes, isotypes etc. I have decided, however, to designate all the type specimens simply as "type", otherwise difficulties are encountered in the designation of say, the holotype, when there are several specimens of the same species. This is best done by a monographer who may need to select a specimen as a lectotype. For the most part, the Thonning specimens at Copenhagen are holotypes of Vahl's, Schumacher's and Thonning's species, while the holotypes of Willdenow's species are Isert's at Berlin, with isotypes at Copenhagen. When there are both Isert and Thonning specimens at Copenhagen I consider Thonning's to be the type, since he presumably described his own specimens, unless stated otherwise. Many of the specimens at Paris,

Leningrad, Geneva, Florence, Stockholm and Munich are probably types but I have not indicated them as such unless there is clear evidence.

Although many of the 63 types reported missing by Junghans have been traced in herbaria other than Copenhagen, the following 36 have so far not come to light anywhere.

Achyranthes nodosa Vahl ex Schum., *A. geminata* Thonning, *Allium guineense* Thonning, *Alternanthera thonningii* Schumacher, *Amaryllis nivea* Thonning, *A. trigona* Thonning, *Andropogon canaliculatus* Schumacher, *A. guineensis* Schumacher, *A. tectorum* Schum. et Thonn., *A. simplex* Vahl et Schum., *A. verticellatus* Schumacher, *Asystasia quaterna* (Thonning) Nees, *Benzonia corymbosa* Schumacher, *Bidens pilosa* Schum. et Thonn., *Buelowia illustris* Schum. et Thonn., *Diodia maritima* Thonning, *Elytraria marginata* Vahl, *Erigeron spathulatum* Schum. et Thonn., *Euphorbia drupifera* Thonning, *Ficus lutea* Vahl, *F. calyptrata* Vahl, *F. microcarpa* Vahl, *F. umbellata* Vahl, *Gossypium punctatum* Schum. et Thonn., *Laschia delicata* Fr., *Loranthus thonningii* Schumacher, *Lundia monacantha* Schum. et Thonn., *Momordica anthelmintica* Schum. et Thonn., *Nymphaea dentata* Schum. et Thonn., *Ocymum sylvaticum* Thonning, *Portulaca prolifera* Schum. et Thonn., *Psychotria? chrysorhiza* Thonning, *Saccharum punctatum* Schumacher, *Sida decagyna* Schum. et Thonn., *Tragia monadelpha* Schum. et Thonn., *Trianthema flexuosa* Schum. et Thonn.

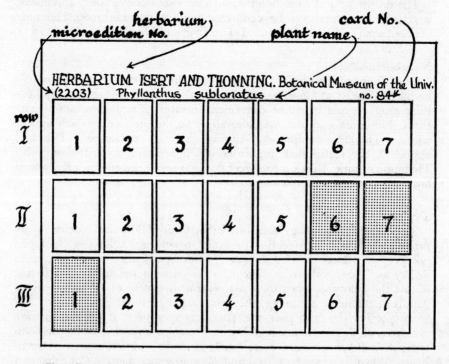

Fig. 1. Example of a microfiche showing the information and the method of numbering individual photographs, e.g. the shaded ones would be cited (in IDC micro-edition No. 2203) as microf. 84: II. 6, 7; III. 1.

Michrofiche

Microfilm editions prepared by the International Documentation Company* exist for most of the sheets. At Copenhagen the whole of the Isert and Thonning herbarium has been filmed (on IDC No 2203), although some 30 sheets have come to light since the photography was completed, and three were included on IDC 2204 "Types in Museum Botanicum Hauniense". Photographs of duplicates are to be found on the IDC 7440 of the Willdenow Herbarium, Berlin; on IDC microedition of the De Candolle Herbarium at Geneva, and on IDC 6201 of the Jussieu Herbarium at Paris. Full lists of these are given as appendices. Plans are being made to film the Webb Herbarium at Florence, too. I am grateful for the interest and co-operation of Mr. Henri L. de Mink, Director of IDC, during the preparation of this work.

Vernacular Names

These have been extracted from Schumacher (1827) and the names may be followed by the language or name of the tribe. A list of 163 vernacular names is included as Appendix I.

Field, Economic and Taxonomic notes

Thonning often provided interesting field data, which he wrote in Danish. Schumacher published them together with his own taxonomic observations in "Beskrivelse", where they have been overlooked by non-Danes since the language is little understood outside Scandinavia. In 1951 W. C. Worsdell translated all the notes into English and I am taking the opportunity of incorporating the translation here since it has lain in MS. at Kew and has never been published. The notes are often followed by the initials of the author "Th." or "S." respectively. Worsdell's translation has been carefully checked by Mrs A. Fox Maule. It should be noted that the translation is rather literal and the botanical terminology of the early nineteenth century has been retained for the taxonomic notes. Ethnobotanists and geographers should find the economic information and vernacular names especially valuable.

The Latin diagnoses and descriptions by both Thonning and Schumacher and published in Schumacher's work are not repeated in the present account.

ACKNOWLEDGEMENTS

This work could not have been completed without the splendid co-operation of the staff at the Botanical Museum, Copenhagen. The present Keeper, Dr Bertel Hansen, kindly facilitated publication by negotiating for a very substantial grant from the Carlsberg Foundation, Copenhagen, for printing which has been undertaken in co-operation with the Bentham-Moxon Trust of the Royal Botanic Gardens, Kew. I am indebted to the Trustees of both the Carlsberg Foundation and the Bentham-Moxon Trust for their generosity. The former Keeper, Dr A. Skovsted, made the Copenhagen herbarium material available, and Mrs Anne Fox Maule has been especially helpful in checking the translation and in many other respects. Dr Jens

* International Microfiche Centre, IDC, Uiterstegracht 45, Leiden, The Netherlands.

Junghans' basic sorting out of the material and his enumeration of the Copenhagen material have also been a great help. Mr J. B. Hall of Ghana has made many useful suggestions and helped with the gazetteer; Dr W. T. Stearn and Dr B. Verdcourt kindly looked over the final MS. Mr J. P. M. Brenan, Deputy Director of Kew, gave encouragement by facilitating visits to other herbaria; I am also grateful to the Directors and Curators of the herbaria in which duplicates have been found, for the loan of material to Kew and for permission to consult their herbaria during visits, and particularly for the aid of Dr C. Steinberg at Florence. I must especially thank Mrs Fiona Neate of Kew for very great assistance in the preparation of the MS. at all stages, and Mrs C. Barndon for helping with the Indexes.

DICOTYLEDONS

ACANTHACEAE

Asystasia quaterna (*Thonning*) *Nees* in A. DC. 1847: 166. *Asystasia coromandeliana* Nees—FTA. 5: 131 (1899), as to synonym. *A. calycina* Benth.— FWTA. ed. 2, 2: 413 (1963), as to synonym.

Ruellia quaterna Thonning in Schum. 1827: 284 and 1829: 58; Junghans 1961: 320.

Exsicc.: "Guinea", *Thonning* 330 (no type specimen traced).

Vernacular name: 'Blaabaa-fye'.

"Frequently found. Used as cabbage etc. Th."
"Differs from *R. intrusa* Vahl only in the number of bracts. S."

Note: Prof. J. K. Morton is convinced that this is a good species, although allied with *A. gangetica* (L.) T. Anders. rather than *A. calycina* as previously thought.—F.N.H.

Barleria opaca (*Vahl*) *Nees*—FWTA. ed. 2, 2: 421 (1963).

Justicia opaca Vahl 1804: 133; Schum. 1827: 10 and 1828: 30; Junghans 1961: 320.

Exsicc.: "Guinea," *Thonning* s.n. (**C**—1 sheet, type, microf. 62: III. 1, 2).

Blepharis maderaspatensis (*L.*) *Heyne ex Roth* subsp. **rubiifolia** (*Schum. et Thonn.*) *Napper* in Kew Bull. 24: 325 (1970). *B. maderaspatensis* (L.) Heyne ex Roth—FWTA. ed. 2, 2: 410 (1963). *B. boerhaviaefolia* Pers.—FTA. 5: 96 (1899).

B. rubiaefolia Schum. et Thonn. in Schum. 1827: 292 and 1829: 66; Junghans 1961: 320.

Exsicc.: "Guinea", *Thonning* 184 (**C**—3 sheets, types, microf. 10: III. 3—6; 11: I. 1; **P–JU**—1 sheet, No. 5724, microf. 408: III. 2).

"Rare. Flowers in July."
"Very like *Acanthus maderaspatensis* but in this latter the internodes are shorter, the leaves ovate, often excisely toothed, but rough with raised points, the outer calyx-leaves are smooth (not white haired) and the flowers occur in clusters in the leaf-axils. S."

Elytraria marginata *Vahl* 1804: 108 ("Habitat ad Senegal, in Guinea. *Thonning*"); FWTA. ed. 2, 2: 418 (1963); Schum. 1827: 9 and 1828: 29 (no MS); Junghans 1961: 320.

Exsicc.: "Guinea", *Thonning* s.n. (no type material traced).

NOTE: Vahl cited a Thonning specimen which would be the holotype if its wherabouts were known. Vahl's mention of Senegal, as well as "Guinea" (i.e. Ghana) seems to be an error, as even now it is not known from the former country.—F.N.H.

Hygrophila abyssinica (*Hochst. ex Nees*) *T. Anders.*—FWTA. ed. 2, 2: 395 (1963).

EXSICC.: "Guinea", *Isert* s.n. (**C**—1 sheet, microf. nil).

NOTE: This specimen has remained unidentified up to now and it was not photographed for the microfiche set.—F.N.H.

Hygrophila auriculata (*Schum. et Thonn.*) *Heine*—FWTA. ed. 2, 2: 395 (1963). *H. spinosa* T. Anders.—FTA. 5: 31 (1899).

Barleria auriculata Schum. et Thonn. in Schum. 1827: 285 and 1829: 59–61; Junghans 1961: 319.

EXSICC.: Ada, 1784, *Isert* s.n. (**C**—1 sheet, microf. 10: II. 5); "Guinea", *Thonning* 185 (**C**—1 sheet, type, microf. 10: II.3, 4).

"Not common; flowers in July".

"Much resembles *B. longifolia* but in the latter the stem is quadrangular not compressed and less hairy; the leaves have a broad stalk and are not furnished at the base with lunate auricles, they are 3 inches long or slightly over (not 6–10 inches); the leaves in the whorls are narrower, lanceolate, much shorter, 1½ inches long by 3 lines broad, and not cordate-lanceolate, near the base an inch broad and near the apex 2 inches long, the corolla is quite smooth not soft haired. For the rest, it approaches another plant which Isert found in Guinea and which Vahl in manuscript termed *Barleria glabrata* and which I have had sent from East Indies by König under the name of *Truxia spinosa*; but this latter differs from our plant in being completely smooth, in having a more quadrangular stem, without flattened sides, in very long sword-shaped leaves; leaves of the whorls cordate-lanceolate with few white marginal hairs; the corollas smooth. Thus *B. glabrata* differs from this one in its smoothness, from *B. longifolia* in the auricled leaves, the cordate-lanceolate whorl-leaves with scattered white marginal hairs, and from *B. auriculata* in the quadrangular stem with somewhat impressed sides, smooth and longer corollas. S."

EXSICC.: "Guinea", *Isert* s.n. (**B**—1 sheet, No. 11658, microf. 829: III. 7).

Justicia flava (*Forsk.*) *Vahl*—FWTA. ed. 2, 2: 428 (1963).

J. plicata Vahl et Thonn. in Vahl 1804: 156 ("Habitat in Aquapim Guineae. *Thonning*"); Schum. 1827: 11 and 1828: 31; Junghans 1961: 320.

EXSICC.: Aquapim, *Thonning* 214 (**C**—1 sheet, type, microf. 62: I. 3, 4).

"Guinea", *Mortensen* s.n. (**C**—1 sheet, microf. 62: I. 5).
"Very common in and at Aquapim".

Phaulopsis ciliata (*Willd.*) *Hepper* in Kew Bull. 28: 328 (1973). *Origanum ciliatum* Willd. 1800b: 133 ("Habitat in Guinea"); Junghans 1961: 340. *P. alcisepela* C. B. Cl.—FWTA. ed. 2, 2: 399 (1963).

EXSICC.: "Guinea", *Isert* s.n. (**B**—1 sheet, No. 10972, type, microf. 777: III. 2).

Acanthaceae indet. in Junghans 1961: 320.

EXSICC.: Whydah, 1785, *Isert* s.n. (**C**—1 sheet, microf. 94: I. 5).

NOTE: No doubt both specimens are from the same gathering made in Dahomey: the Copenhagen sheet bears the MS names *Ruellia perilloides* and *Origanum guineense.*—F.N.H.

Ruspolia hypocrateriformis (*Vahl*) *Milne-Redh.*—FWTA. ed. 2, 2: 431 (1963). *Eranthemum hypocrateriforme* (Vahl) R. Br. ex Roem. & Schult.— FTA. 5: 171 (1899).

Justicia hypocrateriformis Vahl 1804: 165 ("Habitat in Guinea, *Thonning*"); Schum. 1827: 11 and 1828: 31; Junghans 1961: 320.

EXSICC.: "Guinea", *Thonning* 28 (**C**—5 sheets, type, microf. 62: I. 6, 7; II. 1–7); "e Guinea" (**FI**—1 sheet; **P–JU**—1 sheet, No. 5767A, microf. 411: III. 3).

"Grows commonly among bushes on the seashore, but also often inland, where it has a larger and more complete growth. It flowers nearly the whole year through."

"In dry weather and strong sunshine it very easily loses its leaves, but quickly acquires new ones with the first refreshing rain or strong dews. The leaves have a slightly acid taste. Th."

AIZOACEAE (FICOIDACEAE)

Sesuvium portulacastrum (*L.*) *L.*—FWTA. ed. 2, 1: 135 (1954).

S. brevifolium Schum. et Thonn. in Schum. 1827: 233 and 1829: 7, as 'brevieolium' (MS No. 132); Junghans 1961: 338.

EXSICC.: See note below.

VERNACULAR NAME: 'Imbebi'.

"Grows on the side of salt-rivers."

"Much resembles *S. portulacastrum* but in this latter the branches are shorter, less spread out, the leaves shorter stalked longer than the internodes and more pointed. Plucknet's (sic) figure tab. 216, f.1 agrees more closely with the plant here described than the specimens of *S. portulacastrum* I have seen. S."

Ficoidaceae indet. in Junghans 1961: 338.

EXSICC.: "Guinea", *Isert* s.n. (**C**—1 sheet, microf. 98: II. 4).

NOTE: Thonning's description presumably was based on his own specimens which are not at Copenhagen. Isert's specimen agrees very well with the description and I think it is possible that it could have been used by Schumacher, which would therefore make it the type of *s. brevifolium*.—F.N.H.

Trianthema portulacastrum *L.*—FWTA. ed. 2, 1 : 136 (1954).

T. flexuosa Schum. et Thonn. in Schum. 1827: 221 and 1828: 241; Junghans 1961: 338.

EXSICC.: Christiansborg, *Thonning* 66 (no type specimen traced).

"Near Christiansborg; flowers in May and June."
"*T. monogyna* differs from our plant in its stiff branches, segments of 1½ inches long, leaves which are larger, obovate not obcordate, the leaf-stalks longer (½ inch), corolla 5-leaved; the style of same length as corolla. *T. obcordata* which Wallich sent from E. Indies, differs, as far as can be judged from dried specimens, from *T. monogyna* from W. Indies. It may be questionable whether our plant is really a different species from *monogyna*."

Ficoidaceae indet. in Junghans 1961: 338.

EXSICC.: Ada, 1784, *Isert* s.n. (**C**—1 sheet, microf. 107: I. 1).

AMARANTHACEAE

Achyranthes nodosa Vahl ex Schum. 1827: 139 and 1828: 159; Bentham in Hook. 1849: 493; FTA. 6, 1: 65 (1909); FWTA. ed. 2, 1: 152 (1954); Junghans 1961: 320.

EXSICC.: Whydahr, *Isert* (no type specimen traced).

"This plant is found in Isert's herbarium."

NOTE: This remains an imperfectly-known species—F.N.H.

Aerva lanata *(L.) Juss. ex Schult.*—FWTA. ed. 2, 1 : 149 (1954).

Illecebrum lanatum (L.) Murr.—Schum. 1827: 144 and 1828: 164; Junghans 1961: 321.

EXSICC.: Aquapim, *Thonning* 205 (**C**—4 sheets, microf. 55: II. 7; III. 1–7).

"Occurs in Aquapim, but not common."
"Our plant differs only from specimens from East Indies in that the leaves are larger and less hairy. S."

Alternanthera pungens *Kunth*—Melville in Kew Bull. 13: 174 (1958).

Cyathula prostrata (L.) Blume—FTA. 6, 1: 43 (1909). *Alternanthera echinata* Sm.—FTA. 6, 1: 74 (1909). *A. repens* (L.) Link—FWTA. ed. 2, 1: 154 (1954).

Illecebrum obliquum Schum. et Thonn. in Schum. 1827: 142 and 1828: 162; Junghans 1961: 321.

EXSICC.: "Guinea", *Thonning* 193 (**C**—2 sheets, type, microf. 56: I. 1–4).

VERNACULAR NAME: 'Samangkama'.

"Common."

"Dillenius' Figure in Elth. T.7. f.7, which is *Illecebrum achyranthes* does not answer to our plant for the leaves are ovate not cuneate and oblique, and in the figure more pointed. Nor does Willdenow state that the heads are downy. S."

"It is a dangerous weed on footpaths and cultivated places, especially for the natives who go bare foot. Th."

Achyranthes thonningii Schumacher 1827: 139 and 1828: 159; Junghans 1961: 321. *Pupalia thonningii* (Schumacher) Moq.—FTA. 6, 1: 49 (1909).

EXSICC.: Aquapim, *Thonning* 215 (no type specimen traced).

"In copse wood in and at Aquapim."

Alternanthera sessilis (*L.*) *DC.*—FWTA. ed. 2, 1: 154 (1954).

Illecebrum sessile (L.) L.—Schum. 1827: 143 and 1828: 163; Junghans 1961: 321.

EXSICC.: "Guinea", 1784, *Isert* s.n. (**C**—1 sheet, microf. 56: I. 5); Whydah, 1785, *Isert* s.n. (**C**—1 sheet, microf. 56: I. 6); "Guinea", *Thonning* 142 (**C**—2 sheets, microf. 56: I. 7; II. 1–3).

"Grows preferably in moist places, where it is fairly common."

"Plukenet's figure which is cited by Willdenow belongs to *Illecebrum ficoideum*."

Amaranthus lividus *L.*—FWTA. ed. 2, 1: 148 (1954).

Amaranthaceae indet. in Junghans 1961: 321.

Exsicc.: "Semina e Guinea" (**C**—1 sheet, microf. 4: III. 4, 5).

Amaranthus viridis *L.*—FTA. 6, 1: 17 (1909); FWTA. ed. 2, 1: 148 (1954). *Chenopodium caudatum* Jacq. 1788: 325; 1786–93a, t. 344; Junghans 1961: 329.

EXSICC.: "Guinea", *Isert*, s.n. (**B**—2 sheets, No. 5357, type, microf. 365: I. 5, 6).

Amaranthus polystachyus Willd.—Schum. 1827: 407 and 1829: 181; Junghans 1961: 321.

EXSICC.: "Guinea", *Thonning* 267 (**C**—3 sheets, microf. 4: II. 7, III. 1–3; 5: I. 1–2).

VERNACULAR NAMES: 'Maajaa' (of Guinea inhabitants); 'Kuppei Kirei' (in Tamil).

"Here and there."

"The leaves almost exactly resemble spinach when prepared. The natives collect the wild plants. Th."

"The Guinea plant agrees entirely with specimens from East Indies. S."

Celosia argentea *L.*—FTA. 6, 1: 17 (1909); FWTA. ed. 2, 1: 146 (1954).

C. splendens Schum. et Thonn. in Schum. 1827: 140 and 1828: 160; Junhangs 1961: 321.

Exsicc.: Whydah, 1785, *Isert* s.n. (**C**—1 sheet, microf. 17: II. 3); Lahtebierg, *Thonning* 217 (**C**—4 sheets, type, microf. 17: I. 3–7; II. 1–2).

"Rare." "Thonning found it only once near Lahtebierg."

"It differs from the other species such as *C. argentea, albida, margaritacea* in its longer stem, long angle-furrowed thicker flower-stalks; broad-linear, obtuse leaves provided with a mucro, leafy 'bracts', and thicker cylindrical spikes. S."

Celosia trigyna *L.*—FWTA. ed. 2, 1: 146 (1954); Townsend in Hooker's Icones Pl. t. 3704 (1975).

C. laxa Schum. et Thonn. in Schum. 1827: 141 and 1828: 161; Junghans 1961: 321.

Exsicc.: Ada, 1784, *Isert* s.n. (**C**—1 sheet, microf. 17: I. 2); "Guinea", *Thonning* 180 (**C**—1 sheet, type, microf. 16: III. 5–6).

"Common".

"*Celosia trigyna* has smaller leaves, which are more pointed and more obtuse at the base; the leaf-stalks are shorter and closer together, stipulate smaller and more scythe shaped, spike longer, lower flowers farther from each other, calyx more open, ovary of same length as calyx and the style protrudes from the latter. S."

Celosia isertii *C. C. Townsend* in Hooker's Icones Pl.t. 3711 (1975).

C. laxa sensu FTA. 6, 1: 18 (1909); FWTA. ed 2, 1: 147 (1954), *non* Schum. et Thonn.

Exsicc.: Whydah, 1785, *Isert* s.n. (**C**—1 sheet, holotype, microf. 17: I. 1).

Cyathula achyranthoides *(Kunth) Moq.*—FWTA. ed. 2, 1:151 (1954).

Achyranthes geminata Thonning in Schum. 1827: 138 and 1828: 158; Junghans 1961: 320. *Cyathula geminata* Moq.—FTA. 6, 1: 44 (1909).

Exsicc.; Aquapim, *Thonning* 216 (no type specimen traced).

"Grows in woods in and at Aquapim."

Cyathula prostata *(L.) Blume*—FWTA. ed. 2, 1: 149 (1954).

Amaranthaceae indet. in Junghans 1961: 321.

Exsicc.: "Guinea", *Mortensen* s.n. (**C**—1 sheet, microf. 4: III. 6–7).

Pandiaka angustifolia *(Vahl) Hepper* in Kew Bull. 25: 189 (1971). *Achyranthes heudelotii* Moq. in DC. 1849: 310. *A. angustifolia* Benth. (1849). *Pandiaka heudelotii* (Moq.) Hook. f. (1880)—FWTA. ed. 2, 1: 151 (1954).

Gomphrena angustifolia Vahl—Schum. 1827: 157 and 1828: 177; Junghans 1961: 321.

Exsicc.: "Guinea", *Thonning* 186 (**C**—3 sheets, microf. 47: III. 4–7; 48: I. 1, 2; S (1 sheet, type); "e Guinea" (**C**—1 sheet, microf. 48: I. 3, 4; **P–JU** —1 sheet, No. 4590, microf. 313: I. 2).

"Not common; flowers in June and July."

"The Guinea plant differs from the Indian only in having more hairy leaves and twigs, also that the twigs below the spikes have felted-silky hairs, the spikes are larger, thicker, the calices more downy and with longer chaff-scales. S."

NOTE: The type of *Gomphrena angustifolia* Vahl (1794) is supposedly a "Koenig" sheet collected in India, but the genus Pandiaka to which this species really belongs, is not known in India and the sheet was probably collected by Isert since Thonning did not return until 1803.—F.N.H.

Philoxerus vermicularis (*L.*) *P. Beauv.*—FWTA. ed. 2, 1: 153 (1954).

Gomphrena cylindrica Schum. et Thonn. in Schum. 1827: 158 and 1828: 178; Junghans 1961: 321.

Exsicc.: "Guinea", *Thonning* 165 (**C**—1 sheet, type, microf. 48: I. 5, 6).

"Common on the margins of salt lagoons or their bed when they are dried up."

"The authors' description of *G. vermicularis* correctly fits our plant in the dried state but dried specimens of *G. vermicularis* nevertheless have a different appearance from it, and differ in the spathulate-lanceolate light-green leaves of 2 inches in length; the leaf-axils distinctly downy, the spikes ovate not oblong, about 4 lines long, and more ovate globular. Our plant in dried state has linear dark-green leaves of an inch in length, the leaf-axils are devoid of down, the spike cylindric; the flowers more densely imbricate. The habit is also different. S."

Pupalia lappacea (*L.*) *Juss.*—FWTA. ed. 2, 1: 151 (1954).

Achyranthes mollis Thonning in Schum. 1827: 137 and 1828: 157; Junghans 1961: 300.

Exsicc.: "Guinea", *Thonning* 163 (**C**—3 sheets, type, microf. 2: I. 7; II. 1–5); "e Guinea" (**FI**—1 sheet; **P–JU**—1 sheet, No. 4582 B partly, microf. 312: II. 3).

VERNACULAR NAME: 'Mem'lemeté'.

"Common."

"Very like *A. echinata* and perhaps the same plant; only distinguished from it in having herbaceous climbing or creeping stem. S."

"From the whole plant an ash-dye is prepared which natives use for painting blue. Th."

NOTE: Although the Florence sheet is not labelled as being from "Guinea", it is written up as "Achyranthes scandens" by Vahl, the name that appears on the Copenhagen and Paris sheets, which are similar specimens in early flowering state—F.N.H.

AMPELIDACEAE see VITACEAE

ANACARDIACEAE

Sorindeia warneckei *Engl.*—FWTA. ed. 2, 1 : 738 (1958).

Exsicc.: "Guinea", *Thonning* s.n. (**C**—1 sheet, microf. 5: II. 7; also 1 sheet not photographed).

Note: Thonning's notes on the sheet show that he considered this to be a new genus and he named the plant *Anasavi guineensis*, a MS name which has never been taken up.—F.N.H.

Spondias mombin *L.*—FWTA. ed. 2, 1: 728 (1958). *S. lutea* L.—FTA. 1 : 448 (1868).

Exsicc.: Whydah, *Isert* s.n. (**C**—1 sheet, microf. 47: III. 2).

S. aurantiaca Schum. et Thonn. in Schum. 1827: 225 and 1828: 245; Junghans 1961 : 322.

Exsicc.: "Guinea", *Thonning* 83 (no type specimen traced).

Vernacular name: 'Adodomi'.

"Here and there in fields, not altogether common."
'Wonder if different from *S. lutea* Lin. sp. pl. I, p. 613? *Sp. Myrobalanus* in Vahl's herbarium differs from our plant in its leaves which are like those of an ash-tree, leaflets which are narrow-lanceolate, acuminate; the panicle nearly double as long, branches of same filiform, flowers very small in distantly placed clusters. S."
"The fruit has a pleasant wine-sour taste; but quickly affects the teeth. The leaves are boiled in water for a vapour bath against dropsy. Th."

ANNONACEAE

Annona senegalensis *Pers.* subsp. **senegalensis**—Le Thomas in Adansonia, sér. 2, 9: 97 (1969); Verdcourt in FTEA. Annonac. 115, fig. 27 (1971).

A. arenaria Thonning in Schum. 1827: 257 and 1829: 31; FWTA. ed. 2, 1 : 52 (1954); Junghans 1961 : 322.

Exsicc.: Quita, *Thonning* 128 (no type specimen at **C**; **P–JU**—1 sheet, No. 10779 partly, lectotype, microf. 799: II. 3 partly).

Vernacular name: 'Naivié'

"Grows near Quita fairly commonly amongst other shrubs in loose sandy gravel. Flowers in Sept., Oct., Nov. and bears fruit in Dec., Jan. and Feb."
"The fruit has a pleasant but weak odour, a sweet slightly aromatic taste, but only a little pulp. Perhaps with cultivation it will excel others of the same genus. A decoction of the dried leaves is used for old legbone injuries."

NOTE: See Le Thomas (*l.c.*) for a detailed discussion and full synonymy. No doubt the Paris specimen is a duplicate of *Thonning* 128 and was logically designated a lectotype by Le Thomas.—F.N.H.

Annona glauca *Schum. et Thonn.* in Schum. 1827: 259 and 1829: 33; FWTA. ed. 2, 1 : 52 (1954); Junghans 1961 : 322.

EXSICC.: Ursua, *Thonning* 172 (no type specimen at **C**); "e Guinea" (**FI**—1 sheet; **P–JU**—1 sheet, No. 10781, microf. 799: II. 5).

"In neighbourhood of Ursua; flowers in May."
"The fruit has a taste slightly like Guadeloupe melons; it ripens in July. Th."

Annonaceae indet. in Junghans 1961 : 323.

EXSICC.: "Guinea", 1786, *Isert* s.n. (**C**—1 sheet, microf. 109: I. 4).

NOTE: As the description is by Thonning the type must have been collected by him, too, and it is unlikely that Isert's specimen is the type. The specimens in Florence and Paris are almost certainly isotypes of *Thonning* 172—F.N.H.

Uvaria chamae *P. Beauv.*—FTA. 1 : 22 (1868); FWTA. ed. 2, 1 : 36 (1954).

Unona macrocarpa Dunal 1817: 103 ("Hab. in Guinea"); DC. 1818: 489; Junghans 1961 : 322.

EXSICC.: "e Guinea" (**FI**—1 sheet, type?; **P–JU**—1 sheet, No. 10789, type?, microf. 800: I. 2).

Uvaria cylindrica Schum. et Thonn. in Schum. 1827: 256 and 1829: 30; Junghans 1961 : 323.

EXSICC.: Ada, 1784, *Isert* s.n. (**C**—1 sheet, microf. 108: III. 7; "Guinea", *Thonning* 44 (**C**—2 sheets, types, microf. 109: I. 1, 2); "e Guinea" (**FI**—1 sheet).
VERNACULAR NAME: 'Abada'.

"Here and there".
"The leaves have a laurel-like odour; they are pounded fresh and applied to old legbone injuries. The root is used to dispel tumour testiculi; a decoction is used internally and an ointment externally, derived from the finely grated root. The decoction is also drunk for gonorrhoea. The seeds are enveloped in a tasty sweetish jelly, they are roasted and eaten. Th."
"β" "foliis ovato-oblongis breviter acuminatis; acumine obtuso-emarginato".
"I do not venture to determine if this is different [species] from the preceding or a variety. S."

Annonaceae indet. in Junghans 1961 : 323.

EXSICC.: "Guinea", 1786, *Isert* s.n. (**C**—1 sheet, microf. 109: I. 3).

Uvaria ovata (*Vahl ex Dunal*) *A. DC.* subsp. **ovata**—FWTA. ed. 2, 1 : 36 (1954). *Unona ovata* Vahl ex Dunal 1817: 104; DC. 1818: 489; Junghans 1961 : 323. *Uvaria ovata* Vahl ex DC. 1818: 489; Junghans 1961 : 323.

EXSICC.: "e Guinea" (**FI**—1 sheet, type, see note below).

Uvaria cordata Schum. et Thonn. in Schum. 1827: 255 and 1829: 29; FTA. 1: 22 (1868); Junghans 1961: 323.

EXSICC.: "Guinea", *Thonning* 80 (**C**—2 sheets, type, microf. 108: III. 5–6).

VERNACULAR NAME: 'Agingeli'.

"Here and there not rare; flowers in May and June."
"The fruit which contains a sweet mucus, is greedily eaten by the natives. Both the root, wood and bark are used in a decoction for the cure of old legbone injuries. The leaves have a taste slightly like laurel leaves. Th."

NOTE: *Unona ovata* was probably based on a duplicate said by De Candolle to be in Paris (Jussieu), but I have failed to find any specimen in that collection. The Florence specimen under this name may be considered an isotype.— F.N.H.

APOCYNACEAE

Alafia scandens (*Thonning*) *De Wild.*—FWTA. ed. 2, 2: 73 (1963). *Blastotrophe scandens* (Thonning) F. Didr. 1855a: 193. *Alafia landolphioides* (A. DC.) K. Schum.—FTA. 4, 1: 44 (1902).

Nerium scandens Thonning in Schum. 1827: 148 and 1828: 168; Junghans 1961: 323.

EXSICC.: Aquapim, *Thonning* 270 (**C**—4 sheets, types, microf. 69: III. 5–7; 70: I. 1–5).

"Here and there in the woods of Aquapim."

Ancylobotrys scandens (*Schum. et Thonn.*) *Pichon*—FWTA. ed. 2, 2: 60 (1963). *Landolphia scandens* (Schum. et Thonn.) F. Didr. (1855)—FTA. 4, 1: 44 (1902).

Strychnos scandens Schum. et Thonn. in Schum. 1827: 127 and 1828: 147; Junghans 1961: 323.

EXSICC.: "Guinea", *Thonning* 281 (**C**—5 sheets, type, microf. 105: II. 5–7; III. 1–3).

VERNACULAR NAME: 'Abontaa'.

"In Aquapim."
"The natives enjoy the acid mucus which covers the seeds and regard it as wholesome."

NOTE: Junghans deals with a slight confusion over the use by Didrikson and Pichon of the names *Echites guineensis*, *Strychnos guineensis* and *Landolphia thonningii* as supposed synonyms of this species. The last two names are *nomina nuda* of no significance.—F.N.H.

Carissa edulis *Vahl*—FTA. 4, 1: 89 (1902); FWTA. ed. 2, 2: 54 (1963).

Carissa dulcis Schum. et Thonn. in Schum. 1827: 146 and 1828: 166; Junghans 1961: 323.

EXSICC.: "Guinea", *Isert* s.n. (**C**—1 sheet, microf. 15: I. 6); "Guinea",

Thonning 79 (**C**—5 sheets, type, microf. 14: III. 5–7; 15: I. 1–5); "e Guinea" (**FI**—1 sheet; **P–JU**—1 sheet, No. 7198, microf. 529: III. 4).

VERNACULAR NAMES: root 'Akokobessa'; fruit 'Aflaumbe'.

"Fairly common."
"The bark of the root is finely broken up and used as a spice for a dish which bears the name Akokobessa. The berries have a very agreeable taste almost like sweet cherries and afford an excellent broth for sickness. Th."

NOTE: Although Willldenow's Herbarium No. 5107 mentions "Guinea", the specimen is not Isert's.—F.N.H.

Motandra guineensis (*Thonning*) *A. DC.*—FTA. 4, 1: 224 (1902); FWTA. ed. 2, 2: 80 (1963).

Echites guineensis Thonning in Schum. 1827: 149 and 1828: 169; Junghans 1961: 323.

EXSICC.: Aquapim, *Thonning* 262 (**C**—4 sheets, type, microf. 37: III. 7; 38: I. 1–7, II. 1); "e Guinea" (**P–JU**—1 sheet, No. 7118, microf. 523: III. 2).

"Grows in Aquapim."

Rauvolfia vomitoria Afzel.—FWTA. ed. 2, 2: 69 (1963).

Apocynaceae indet. in Junghans 1961: 324.

EXSICC.: Sawi, 1785, *Isert* s.n. (**C**—1 sheet, microf. 91: III. 1).

Voacanga africana Stapf—FWTA. ed. 2, 2: 67 (1963).

Apocynaceae indet. in Junghans 1961: 324.

EXSICC.: "Guinea", *Thonning* s.n. (**C**—1 sheet, microf. 110: I. 5).

ASCLEPIADACEAE*

Calotropis procera (*Ait.*) *Ait. f.*—FTA. 4, 1: 294 (1902); FWTA. ed. 2, 2: 91 (1963).

Asclepias procera Ait.—Schum. 1827: 154 and 1828: 174; Junghans 1961: 324.

EXSICC.: La and Prampram, *Thonning* 154 (**C**—3 sheets, microf. 8: I. 2–7).

"Grows in sandy places at La and Prampram. Flowers in March and April."
"It has a great resemblance to *Asclepias gigantea*, but has larger and broader leaves; the flower-stalks are half as short as the leaves, corolla-lobes are ovate concave erect and inclined towards each other, and contrariwise lanceolate-linear, bent backwards; the nectaries are also smaller. Plukenet's figure belongs rather to *A. gigantea*: and Alpini's figure probably represents another plant, as it has a racemus terminalis. S."

Gymnema sylvestre (*Retz.*) *Schultes*—FTA. 4, 1: 413 (1903); FWTA. ed. 2, 2: 95 (1963).

* Including Periplocaceae.

Cynanchum subvolubile Schum. et Thonn. in Schum. 1827: 150 and 1828: 170;
   Junghans 1961: 324.

Exsicc.: Ada or Blegusso, *Thonning* 308 (**C**—4 sheets, type, microf. 30: I.
3–7, II. 1–2; **S**—1 sheet, type); "e Guinea" (**FI**—1 sheet; **P–JU**—1 sheet,
No. 7033, microf. 516: III. 6).

   "At Ada, Blegusso; not very common."

Leptadenia hastata (*Pers.*) *Decne.*—FWTA. ed. 2, 2: 98 (1963). *L.
lancifolia* (Schum. et Thonn.) Decne.—FTA. 4, 1: 430 (1903).

Cynanchum lancifolium Schum. et Thonn. in Schum. 1827: 150 and 1828: 170;
   Junghans 1961: 324.

Exsicc.: "Guinea" coast, *Thonning* 251 (**C**—6 sheets, types, microf. 29: I.
5–7, II. 1–7, III. 1–2, 4–5; **S**—1 sheet, type); "e Guinea" (**P–JU**—1 sheet,
No. 7057, microf. 518: III. 3).

   "Here and there among the bushes in the neighbourhood of the shore."
   "*Cynanchum reticulatum* Retz. Fasc. 2 p. 15 No. 38 has smaller flowers, ovate
   calyx lobes, not linear-lanceolate, the corollas are shorter, less tomentose,
   the lower leaves ovate-cordate with more oblique nerves. The upper part
   of the branches is so much like our plant that by a hasty look you might
   regard it as the same species."

C. scabrum Schum. et Thonn. in Schum. 1827: 152 and 1828: 172 (no MS
   number); Junghans 1961: 324.

Exsicc.: Whydah, 1785, *Isert* s.n. (**C**—1 sheet, microf. 29: III. 3. This sheet
was cited by Junghans *l.c.* under *C. lancifolium*); "Guinea", *Thonning* s.n.
(**C**—2 sheets, type, microf. 29: III. 6–7; 30: I. 1, 2; **S**—2 sheets, ? types).

   "Here and there among the bushes."

Note: One of the two sheets at Stockholm is labelled only "dedit Thonning".
The specimen does not exactly match the other one so it may not be a type—
F.N.H.

Oxystelma bornuense *R. Br.*—FWTA. ed. 2, 2: 90 (1963).

Asclepiadaceae indet. in Junghans 1961: 324.

Exsicc.: Popo, 1785, *Isert* s.n. (**C**—1 sheet, microf. 78: III. 5).

Pergularia daemia (*Forsk.*) *Chiov.*—FWTA. ed. 2, 2: 90 (1963). *Daemia
extensa* R. Br.—FTA. 4, 1: 387 (1903).

Asclepias convolvulacea Willd. 1798a: 1269 ("Habitat in Guinea"); Schum.
   1827: 152 and 1828: 172; Junghans 1961: 324.

Exsicc.: "Guinea", *Isert* s.n. (**B**—1 sheet, No. 5271, type, microf. 360: I. 3);
"Guinea", *Thonning* 148 (**C**—4 sheets, microf. 7: I. 2–7, II. 1–2); "e Guinea"
(**C**—1 sheet 7: II. 3–4; **FI**—1 sheet; **P–JU**—1 sheet, No. 7061, microf.
518: III. 7).

Vernacular name: 'Kah-Ba'.

   "Common."

"On this specimen in Vahl's herbarium the follicles are stiffly bent backwards. S."

Asclepias muricata Schum. et Thonn. in Schum. 1827: 153 and 1828: 173; Junghans 1961: 324.

Exsicc.: Aquapim, *Thonning* 200 (C—5 sheets, type, microf. 7: II. 5–7, III. 1–7).

Vernacular name: 'Kah-Ba'.

"Frequent in Aquapim among bushes in formerly cultivated places. Flowers in June and July."

"At first glance it resembles *A. convolvulacea* but differs from it in having more oblong pointed leaves, which on both sides are downy-haired and not felted; the calyx is nearly smooth, the corolla quite white and a little larger; tip of nectary bifid, follicles horizontal extended and densely beset with spikes; finally the whole plant is a little larger. Th."

Raphionacme brownii *Scott Elliot*—FWTA. ed. 2, 2: 85 (1963).

Asclepiadaceae indet. of Junghans 1961: 324.

Exsicc.: "Guinea", *Thonning* s.n. (C—1 sheet, microf. 18: II. 1).

Sarcostemma viminale *(L.) R. Br.*—FTA. 4, 1: 384 (1903); FWTA. ed. 2, 2: 93 (1963).

Asclepias nuda Schum. et Thonn. in Schum. 1827: 155, and 1828: 175.

Exsicc.: "Guinea", *Thonning* 303 (C—1 sheet, type, microf. 8: I. 1).

Vernacular name: 'Enkafo'.

"Here and there amongst the bushes, especially near the coast."

"In *Ascl. aphylla* the branches are opposite often dichotomously divided and opposite the two umbels which are situated together; the corollas are smaller and the margins of the lobes flat. *As. stipitacea* Forsk. Aegypt. Cent. 2 p. 50 No. 69 differs in having erect sprawling and bush-like stems, and umbels situated both terminal and in the leaf-axils. S."

Asclepiadaceae indet.—Junghans 1961: 324.

Exsicc.: "Guinea", *Thonning* s.n. (C—1 sheet, microf. 78: III. 6, 7).

Note: Perhaps this is *Oxystelma bornuense* (q.v.) but the specimen has only one leaf attached to a short length of stem.—F.N.H.

BALANOPHORACEAE

Thonningia sanguinea *Vahl* 1810: 125, pl. 6 ("Habitat in sylvis umbrosis montium Aquapim Guineae"); Schum. 1827: 431 and 1829: 205; FTA. 6, 1: 438; FWTA. ed. 2, 1: 667 (1958); Bullock in Kew Bull. 3: 363 (1948).

Exsicc.: Aquapim, *Thonning* 94 (**BM**—1 sheet, type; **C**—2 sheets, types, microf. 106: III. 4–6 also coloured drawings; **K**—1 sheet, type). See note in Introduction p. 9.

"Grows in high shady woods down below and upon the Aquapim mountains."

"A decoction of this plant is used for washing out venereal sores, especially venereal eruptions. It is also used for heightening the colour of parrots' red tail feathers; the old feathers are pulled out and the sore place is rubbed in with the finely grated plant. These feathers are in much use for finery in dress, and their value is fixed according to the colour. Th."

NOTE: The coloured drawings on the two sheets attributed to Vahl by Junghans are more likely to have been done by Thonning who collected the plant and must have known how the specimens change colour on drying. There are six drawings of the plant, as well as some details of the individual flowers. The level of the earth is indicated by the dotted lines, but there is no writing on the sheets. Mrs Fox Maule informs me that "the table in Vahl's paper is marked T. del. M.sc." which indicates that Thonning drew the pictures which were engraved for publication by whoever "M." might be.— F.N.H.

BIGNONIACEAE

Newbouldia laevis (*P. Beauv.*) *Seemann ex Bureau*—FTA. 4, 2: 521 (1906); FWTA. ed. 2, 2: 388 (1963).

Bignonia glandulosa Schum. et Thonn. in Schum. 1827: 274 and 1829: 48; Junghans 1961: 325.

EXSICC.: Whydah, 1785, *Isert* s.n. (**C**—1 sheet, microf. 10: III. 1); "Guinea", *Thonning* 123 (**C**—1 sheet, type, microf. 10. II. 6, 7).

VERNACULAR NAME: 'Nabaa-di'.

"Cultivated here and there."

"Used in the fetish-cult."

"These 2 species [i.e. *B. glandulosa* and *B. tulipifera*] and *B. spathacea* should on account of the divergent appearance of the calyx be separated from the genus and referred to Spathodea Persoon. S."

Spathodea campanulata *P. Beauv.*—FTA. 4, 2: 529 (1906); FWTA. ed. 2, 2: 386 (1963).

Bignonia tulipifera Thonning in Schum. 1827: 273 and 1829: 47; Junghans 1961: 325.

EXSICC.: Aquapim, *Thonning* 95 (**C**—1 sheet, type, microf. 10: III. 2); "e Guinea" (**P-JU**—1 sheet, No. 4954, microf. 342: II. 2).

VERNACULAR NAME: 'Osisiu'.

"In hilly areas of Aquapim and near Frederiksstad. Flowers at different times of the year."

"The bark is used by the natives in dysentery. The flowers are as large as the largest tulips."

BOMBACACEAE

Adansonia digitata *L.*—Schum. 1827: 300 and 1829: 74; FWTA. ed. 2, 1: 334 (1958); Junghans 1961: 325.

Exsicc.: *Thonning* 54 (no material traced).

Vernacular name: 'Sjadjo-tjo'.

"At Quitta, Tubreku, Aquapim."
"The inhabitants in Ashanti, Akim, Aquapim always bury their principal dead secretly and often in this tree especially in times of war, when they fear that the enemy will discover the body and keep the bones on his drums as a sign of victory and of his enemies' disgrace. The inhabitants declare that the body is dried without putrefaction in this tree; at old age the trunk is hollowed out and the body is lowered into it. The wood is loose and useless even for burning. The mealy acid substance which surrounds the seed is eaten by the natives. The whole fruit is burnt to an ash and used with boiled palm-oil for making soap. Th."

Ceiba pentandra *(L.) Gaertn.*—FWTA. ed. 2, 1: 335 (1958). *Eriodendron anfractuosum* DC.—FTA. 1: 214 (1868).

Bombax pentandrum L.—Schum. 1827: 301 and 1829: 75; Junghans 1961: 325.

Exsicc.: *Thonning* 167 (no material traced).

Vernacular name: 'Onjai-tjo'.

Bombax guineense Thonning in Schum. 1827: 302 and 1829: 76; Junghans 1961: 325.

Exsicc.: "Guinea," *Thonning* 160 (C—1 sheet, type, microf. 11: I. 6, 7).

Vernacular name: 'Odum-tjo'.

BORAGINACEAE

Coldenia procumbens *L.*—Schum. 1827: 85 and 1828: 105; FWTA. ed. 2, 2: 321 (1963); Junghans 1961: 325.

Exsicc.: "Guinea", *Thonning* 249 (C—4 sheets, microf. 22: I. 3–6); "e Guinea" (FI—1 sheet; P–JU—1 sheet, No. 6564, microf. 476: II. 3).

"Grows preferably in damp places although not common."
"The Guinea plant is larger in all parts than the Indian".

Cordia guineensis *Thonning* in Schum. 1827: 128 and 1828: 148; FWTA. ed. 2, 2: 320 (1963); Junghans 1961: 325.

Exsicc.: Quitta, 1784, *Isert* s.n. (C—1 sheet, microf. 27: I. 2, 3); "Guinea" coast, *Thonning* 268 (no type material at C; G–DC—1 sheet, 9: 480. 44, type, microf. 1665: I. 8); "e Guinea" (P–JU—1 sheet, No. 6494, microf. 470: III. 1).

VERNACULAR NAME: 'Jumo-saa'.

"Common in shore landscapes; flowers in the rainy period".

Ehretia corymbosa *Bojer ex A. DC.*—Fosberg in Kew Bull. 29: 260 (1974).

E. cymosa Thonning in Schum. 1827: 129 and 1828: 149; FWTA. ed. 2, 2: 318 (1963); Junghans 1961: 326, non Willd. ex Roem. et Schult.

EXSICC.: Ada, 1784, *Isert* s.n. (**C**—3 sheets, microf. 39: I. 5, 6; II. 2, 3); "Guinea", *Thonning* 89 (**C**—4 sheets, type, microf. 38: III. 5–7; 39: I. 1–4, 7; II. 1); "e Guinea" (**FI**—1 sheet; **P–JU**—1 sheet, No. 6504, microf. 471: II. 7).

VERNACULAR NAME: 'Lavasá'.

"Common; flowers in May."
"The longest branches are used as anchor-bands. The natives chew the wood along with the seeds of *Sterculia verticillata* from which is produced a red colour which is used for fetish-ornaments, amulets etc. The flowers smell like *Galium verum*".

Heliotropium indicum *L.*—FWTA. ed. 2, 2: 321 (1963).

Heliotropium africanum Schum. et Thonn. in Schum. 1827: 87 and 1828: 107; FTA. 4, 2: 43 (1905); Junghans 1961: 326.

EXSICC.: "Guinea", *Thonning* 69 (**C**—1 sheet, type, microf. 52: I. 5).

"Common."
"*Heliotropium indicum* which in habit comes rather near our plant, differs in that the stem, leaves, flower-stalks, rachis, calyx and even the corolla-tube are beset with dense white hairs, that the leaves are mostly cordate, sinuous at the base and more wrinkled, tube of corolla less swollen and not double as long as the calyx, and the lobes of the corolla broader".

Heliotropium strigosum *Willd.* 1798a: 743 ("Habitat in Guinea"); Schum. 1827: 86 and 1828: 106; FWTA. ed. 2, 2: 322; Junghans 1961: 326.

EXSICC.: "Guinea", *Isert* s.n. (**B**—1 sheet, No. 3253, type, microf. 219: II. 7; **C**—1 sheet, type if Isert's, microf. 52: II. 6, 7); "Guinea", *Thonning* 56 (**C**—3 sheets, microf. 52: I. 7; II. 2–5); "e Guinea" (**C**—2 sheets, microf. 52: I. 6; II. 1; **FI**—1 sheet; **P–JU**—1 sheet, No. 6571, microf. 477: I. 6).

"Common. Flowers in May, June, and Sept."
"In investigating the plant in its habitat Thonning found that the stem is subherbaceous and ascending, not shrubby and erect, the spikes situated singly not in pairs as Wil(l)denow assumed".

CACTACEAE

Opuntia sp.

Cactus tuna L.—Schum. 1827: 229 (no MS); Junghans 1961: 326.

EXSICC.: (no material traced).

(a) Isert's. (b) Thonning's, including signature at the end of a letter. (c) Schumacher's on the reverse of a Thonning sheet, showing number. (d) Reverse of a Thonning sheet showing Schumacher's amendments. (e) Gay's on a sheet at Kew.

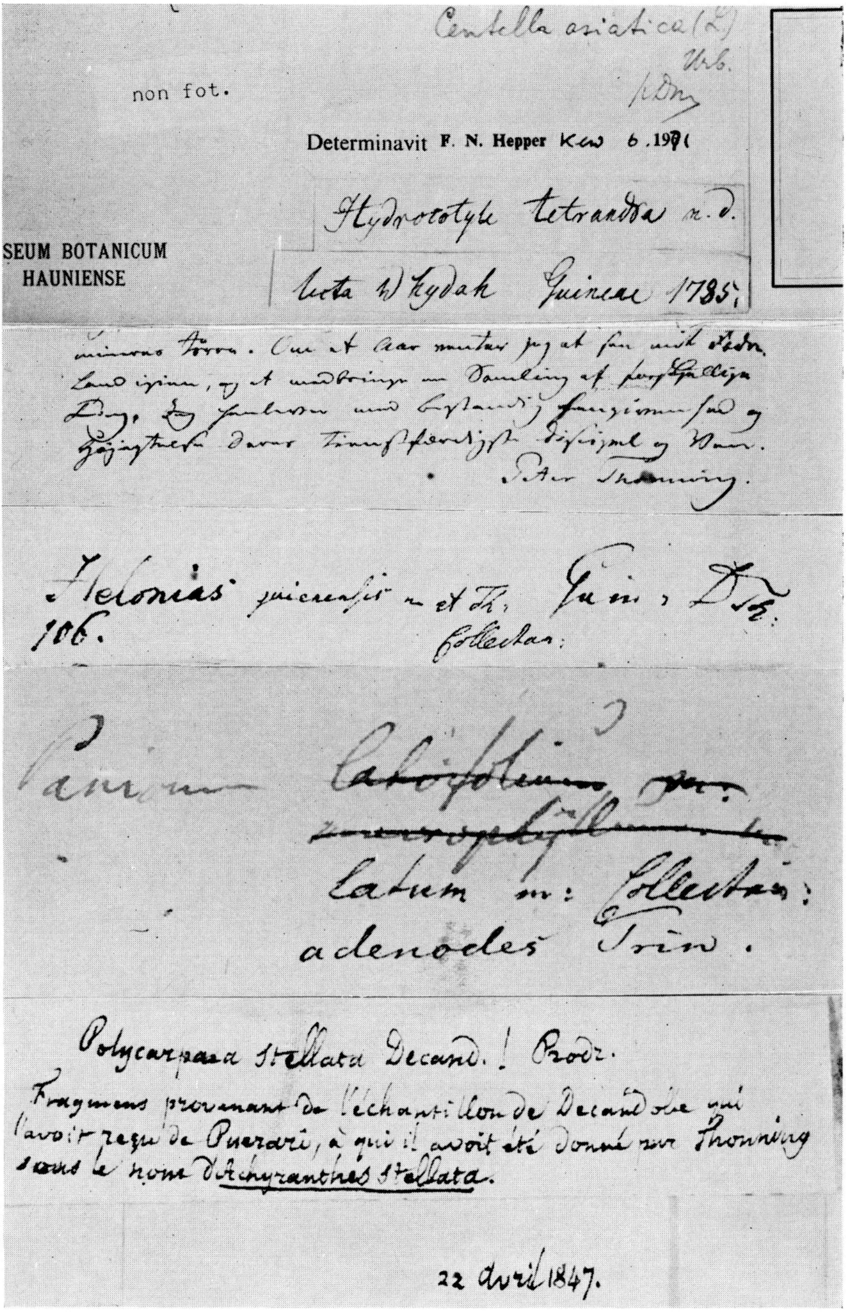

PLATE 2

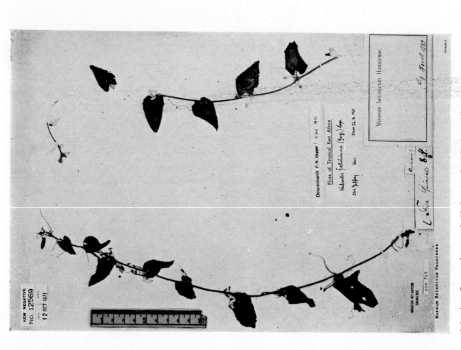

(b) Drawing of *Thonningia sanguinea* Vahl.

(a) An Isert sheet of *Kedrostis foetidissima* (Jacq.) Cogn. at Copenhagen.

PLATE 3

"Cultivated."

NOTE: The brief description indicates that this is an Opuntia, which is a well-known introduction from America. There is only one species at present naturalised along the shores of Ghana—*O. dillenii* (Ker-Gawl.) Haw.— F.N.H.

CAESALPINIACEAE

Afzelia parviflora (*Vahl*) *Hepper* in Kew Bull. 26: 565 (1972).

Westia parviflora Vahl 1818: 119; Oliver in FTA. 2: 294 (1871), note; Keay in Kew Bull. 9: 271 (1954).

EXSICC.: Sierra Leone, 1785, *Smeathman* s.n. (**C**—1 sheet, holotype, microf. 111: I. 6, 7; **BM**—1 sheet, isotype).

NOTE: The identity of *W. parviflora* has for long remained a mystery, which is now settled. The specimen does not belong to this collection of Danish Guinea plants, but it is included here since it appears on microfiche No 2203. —F.N.H.

Berlinia grandiflora (*Vahl*) *Hutch. et Dalz.*—FWTA. ed. 2, 1: 470 (1958).

Westia grandiflora Vahl 1818: 118 ("Habitat in Guinea *Isert*"); Junghans 1961: 327.

EXSICC: Whydah, 1785, *Isert* s.n. (**C**—1 sheet, type, microf. 111: I. 5).

NOTE: Oliver's note on this species in FTA. 3: 294 is superseded by Keay's fuller discussion in Kew Bull. 9: 267 ff. (1954).—F.N.H.

Caesalpinia bonduc (*L.*) *Roxb.*—FWTA. ed. 2, 1: 481 (1958).

Guilandina bonducella L.—Schum. 1827: 210 and 1828: 230; Junghans 1961: 327.

EXSICC.: Ningo, *Thonning* 232 (no material traced).

VERNACULAR NAME: 'Demi-tjo'; seed 'Vualé-mi'.

"Here and there, at Ningo common among Cactus Tuna".
"The seeds are used as playing-stones in a game called Vualé, hence the name 'Vualé-mi'. They are also sometimes used as fetish for hanging on small children. Th.'

Caesalpinia pulcherrima (*L.*) *Sw.*—Schum. 1827: 209 and 1828: 229; FWTA. ed. 2, 1: 481 (1958); Junghans 1961: 326.

EXSICC.: Whydah, 1785, *Isert* s.n. (**C**—1 sheet, microf. 13: III. 1); "Guinea", *Thonning* 60 (**C**—1 sheet, microf. 13: III. 2); "e Guinea" (**C**—1 sheet, microf. 13: II. 6, 7).

"It flowers the whole year through. Is cultivated in Danish establishments on account of its beautiful flowers".

B

Cassia absus *L.*—FTA. 2: 279 (1871); FWTA. ed. 2, 1: 453 (1958).
C. thonningii DC. 1825a: 500; Junghans 1961: 327.

C. viscosa Schum. et Thonn. in Schum. 1827: 205 and 1828: 225.

Exsicc.: Ada, 1784, *Isert* s.n. (**C**—1 sheet, microf. 16: I. 5–6); "Guinea", *Thonning* 63 (**C**—3 sheets, type, microf. 15: III. 5–6; 16: I. 1–4); "e Guinea" (**FI**—1 sheet; **G–DC**—1 sheet, 2: 500. 128, microf. 433: I. 8; **P–JU**—1 sheet, No. 14532, microf. 1068: I. 7).

"Common except in the months of Dec., Jan., Feb."
"Very like *C. absus* but in this the leaves are rounder and more curved, the stipules almost bristle-like, the racemes few-flowered, the hairs on the whole plant much shorter, but on the pods longer, gland-bearing at the base. S".

Cassia mimosoides *L.*—FTA. 2: 280 (1871); FWTA. ed. 2, 1: 452 (1958).
C. microphylla Willd. *β guineensis* DC. 1825: 505.

C. geminata Vahl ex Schum. et Thonn. in Schum. 1827: 208 and 1828: 228; Junghans 1961: 326.

Exsicc.: Ada, 1785, *Isert* s.n. (**C**—1 sheet, microf. 15: III. 4); "Guinea", *Thonning* 49 (**C**—3 sheets, type, microf. 15: II. 3, 4, 7; III. 1–3; **G–DC**— 1 sheet, 2: 505. 182 *β*, type, microf. 434: II. 4); "e Guinea mis Dr. Pflug" (**C**—1 sheet, microf. 15: II. 5, 6).

"Grows commonly in open fields near the shore. Flowers in the rainy season".
"Comes near *C. mimosoides* but is different from it. S."

Cassia obtusifolia *L.*—Brenan in FTEA. Legumin.–Caesalp. 77 (1967).

C. tora sensu Schum. 1827: 204 and 1828: 224; FWTA. ed. 2, 1: 455 (1958); Junghans 1961: 327.

Exsicc.: Ada, *Thonning* 294 (no material traced).

"Here and there, in quantity at Ada, preferably in and near towns".

Cassia occidentalis *L.*—FWTA. ed. 2, 1: 455 (1958).

C. planisiliqua L.—Schum. 1827: 206 and 1828: 226; Junghans 1961: 326.

Exsicc.: "Guinea", *Thonning* 12 (**C**—1 sheet, microf. 15: II. 1–2).

Vernacular name: 'Bãsissa'

"Grows everywhere especially near towns and cultivated spots".
"Is used by the natives in different ways, e.g. the bark of the root is scraped off, boiled with pytto (native beer) and is drunk in cases of dysentery. The same bark is finely macerated along with a few grains-of-Paradise [Aframomum] and with aid of lemon-juice is made into an ointment with which ringworm is coated over. The leaves are used to induce the opening of bowels and to soothe pains in the abdomen and are boiled for this purpose with pytto. The bark of the root has a bitter slightly astringent taste, it is said to be recommended as a good substitute for china-bark [Chinchona]. The leaves have a stupefying odour, which much resembles opium. Th."

Cassia podocarpa *Guill. et Perr.*—FWTA. ed. 2, 2 : 452 (1958).

Caesalpiniaceae indet. in Junghans 1961 : 327.

Exsicc.: Whydah, 1785, *Isert* s.n. (**C**—1 sheet, microf. 15 : I. 7).

Cynometra vogelii *Hook. f.*—FWTA. ed. 2, 2 : 458 (1958).

Caesalpiniaceae indet. in Junghans 1961 : 327.

Exsicc.: prob. *Isert* s.n. (**C**—1 sheet, microf. 30: II. 3–4); "e Guinea" (**C**—1 sheet, microf. 30: II. 5–6).

Dialium guineense *Willd.* 1796: 30–32, pl. 6; 1797a: 49 ("Isert"); FTA. 2: 283 (1871); FWTA. ed. 2, 1 : 449; Junghans 1961 : 327.

Exsicc.: Acra, *Isert* s.n. (**B**—2 sheets, No. 147, type, microf. 9: III. 3, 4; **C**—2 sheets, types, microf. 35: III. 4, 5).

Codarium nitidum Sol. ex Vahl 1804: 302 ("Habitat in Guinea. *Isert, Thonning*"); Schum. 1827: 18 and 1828: 38.

Exsicc.: Acra, *Thonning* 105 (**C**—2 sheets, type, microf. 21: III. 7; 22; I. 1, 2); "e Guinea" (**C**—1 sheet, microf. 35: III, 6–7).

Vernacular name: 'Joj-tjo'; European name: 'Black tamarind'.

"Grows in Acra. Flowers Sept. and May".
"The mealy pulp of the fruit has a pleasant acidity and soaked in water is a very medicinal drink for fever patients. The wood yields good charcoal for the forge and for other uses. Th."

Griffonia simplicifolia (*Vahl ex DC.*) *Baill.*—FWTA. ed. 2, 1 : 446 (1958). *Bandeiraea simplicifolia* (Vahl ex DC.) Benth.—FTA. 2 : 285 (1871).

Schotia simplicifolia Vahl ex DC. 1825a: 508; Schum. 1827: 212 and 1828: 232; Junghans 1961: 327.

Exsicc.: Christianborg and Aquapim, *Thonning* 96 (**C**—4 sheets, type, microf. 95: III. 2–7-microf. card 96 incorrectly headed *Schotia simplicifolia*); "e Guinea" (**FI**—1 sheet; **P–JU**—1 sheet, No. 14622, microf. 1074: III. 1).

"Rare, near to Christiansborg and in Aquapim".

Piliostigma thonningii (*Schumacher*) *Milne-Redhead*—FWTA. ed. 2, 1: 444 (1958). *Bauhinia articulata* sensu FTA. 2: 290 non *B. reticulata* DC.

Bauhinia thonningii Schumacher 1827: 203 and 1828: 223; Junghans 1961: 326.

Exsicc.: Aquapim, *Thonning* 352 (**C**—2 sheets, types, microf. nil; **S**—1 sheet, type).

"At Aquapim".
"The description is from a dried specimen. S."

CAPPARACEAE

Capparis brassii *DC.*—FZ. 1: 237 (1960).

C. thonningii Schumacher 1827: 236 and 1829: 10; FWTA. ed. 2, 1: 90 (1954);
Junghans 1961: 328.

EXSICC: Whydah, *Isert* s.n. (**C**—1 sheet, microf. 14: II. 4); "Guinea",
Thonning 234 (**C**—3 sheets, types, microf. 14: I. 7, II. 1–3); "e Guinea"
(**P–JU**—1 sheet No. 11258, microf. 834: I. 1).

VERNACULAR NAME: 'Otiobibomo'.

"Not common".
"Resembles *C. sepiaria* but in this latter the branches are somewhat
curved, the leaves ovate, nearly pointed, more membranous, almost sinuate
at the base and shorter stalked; the thorns are longer, black not yellowish;
all the flowers form a single umbel which is situated at the top. S".

Capparis erythrocarpos *Isert* 1789: 334, pl. 9, f.3; Willd. 1800a: 1132;
Schum. 1827: 235 and 1829: 9; FWTA. ed. 2, 1: 89 (1954); Junghans 1961:
328.

EXSICC.: Accra, 1786, *Isert* s.n. (**B**—1 sheet, No. 10039, type, microf. 698:
II. 6; **C**—1 sheet, type, microf. 14: I. 3); "Guinea", *Thonning* 99 (**C**—2
sheets, microf. 14: I. 1, 2; **P–JU**—1 sheet No. 11273, microf. 834: III. 6).

VERNACULAR NAME: 'Petipeti'; fruit 'Abaumba'.

"Here and there at seaside".

Capparis tomentosa *Lam.*—Schum. 1827: 234 and 1829: 8; FWTA. ed.
2, 1: 90; Junghans 1961: 328.

EXSICC.: Ningo, *Thonning* 233 (no material traced).

VERNACULAR NAME: 'Petipeti'.

"Rare; only found at Ningo".

Cleome rutidosperma *DC.*—FTEA. Cappar. 11 (1964).

C. ciliata Schum. et Thonn. in Schum. 1827: 294 and 1829: 68; FWTA. ed. 2,
1: 87 (1954); Junghans 1961: 328.

EXSICC.: Ada, 1784, *Isert* s.n. (**C**—1 sheet, microf. 21: I. 5); "Guinea"
Thonning 115 (**C**—2 sheets, type, microf. 21: I. 1, 2).

"Grows here and there, flowers in the rainy season".
"*Cleome triphylla* which resembles it in habit differs in having a smooth
stem, nearly ovate, more obtuse smooth (not marginally-hairy or hairy)
leaves and very short leaf-stalks which are shorter than the leaflets. In my
herbarium is a plant sent from East India by König under the name of
Meyera triphylla which is exactly like the Guinea plant here described. S".

NOTE: The König sheet mentioned is included in the Isert & Thonning
microfiche set on 21: I. 3, 4.—F.N.H.

Crateva adansonii *DC.*—FTEA. Cappar. 20 (1964). *C. religiosa* Forst.
f.—FTA. 1: 99 (1868); FWTA. ed. 2, 1: 90 (1954).

Crataeva guineensis Schum. et Thonn. in Schum. 1827: 240 and 1829: 14; Junghans 1961 : 328.

EXSICC.: R. Volta, *Thonning* 26 (no material traced).

"On the side of the river Volta, flowers in May".

Euadenia trifoliolata *(Vahl ex Thonning) Oliv.*—FTA. 1: 91 (1868); FWTA. ed. 2, 1 : 93 (1954).

Stroemia trifoliata Vahl ex Thonning in Schum. 1827: 114 and 1828: 134 (MS No. 171); Junghans 1961 : 328.

EXSICC.: Aquapim, 1786, *Isert* s.n. (**C**—1 sheet, type, microf. 105: II. 4).

"Thonning found only a single tree in the native place Adau in Aquapim; it flowered in May".

NOTE: There is little doubt that as Vahl provided the name and diagnosis the Isert specimen is the holotype. It is not known whether Thonning even collected material of the tree he found.—F.N.H.

Gynandropsis gynandra *(L.) Briq.*—FTA. 1: 82 (1868); FWTA. ed. 2, 1 : 88 (1954).

Cleome acuta Schum. et Thonn. in Schum. 1827: 293 and 1829: 67; Junghans 1961 : 328.

EXSICC.: "Guinea", *Thonning* 67 (**C**—1 sheet, type, microf. 21 : I. 6).

VERNACULAR NAME: 'Taeta-fye'.

"Here and there, comes out especially in the fertile period of the year".
"The leaves are used as cabbage by the native. Th."
"Resembles *C. pentaphylla* but in the latter the leaves are smaller, the cauline leaves obovate obtuse, the pods provided with short hairs and not with raised points. S."

Ritchiea reflexa *(Thonning) Gilg et Benedict*—FTA. 1 : 98 (1868); FWTA. ed. 2, 1 : 92 (1954).

Capparis reflexa Thonning in Schum. 1827: 237 and 1829: 11; Junghans 1961: 328.

EXSICC.: Whydah, 1785, *Isert* s.n. (**C**—1 sheet, microf. 14: I. 4); "Guinea", *Thonning* 100 (**C**—2 sheets, type, microf. 14: I. 5, 6).

VERNACULAR NAME: 'Ajilebi'.

"Here and there in shore areas".

NOTE: The sheet in **P–JU** (No. 11272, microf. 834: III. 5) distributed by Vahl as *Capparis mucronata* is not *Ritchiea reflexa*; it remains indeterminate.—F.N.H.

CARYOPHYLLACEAE

Polycarpaea eriantha *Hochst. ex A. Rich.* var. **effusa** *(Oliv.) Turrill*—

FWTA. ed. 2, 1 : 759 (1958). *P. corymbosa* (L.) Lam. var. *effusa* Oliv.—FWTA. ed. 2, 1 : 132 (1954). *Achyranthes stellata* sensu Junghans 1961 : 321, partly.

Exsicc.: "Guinea", *Isert* s.n. (**C**—1 sheet, microf. 69: II. 5, 6; **S**—1 sheet).

Polycarpaea stellata (*Willd.*) *DC.* 1828: 374; FTA. 1 : 145 (1868); FWTA. ed. 2, 1 : 132 (1954). *Achyranthes stellata* Willd. 1798a: 1195; Junghans 1961 : 321, partly.

Mollia stellata (Willd.) Willd. 1806b: 11, t. 11; Schum. 1827: 136 and 1828: 156; Junghans 1961 : 321.

Exsicc.: "Guinea", *Isert* s.n. (**B**—1 sheet, No. 4997, microf. 342: III. 2); "Guinea", *Thonning* 243 (**C**—1 sheet, microf. 69: II. 4; **K**—1 sheet, ex Herb. Gay); "e Guinea" (**FI**—1 sheet; **G-DC**—1 sheet ex Herb. Puerari, 3: 374. 7, microf. 556: II. 8; **P-JU**—1 sheet, No. 13367, microf. 989: III. 4).

"Common in high and dry places at all seasons".

"Although Wildenow's description differs from Thonning's, I am quite sure that it is the same plant. *Polycarpaea* Lamarck and *Hagea* Pers. Syn. 1. p. 262 are the same genus as *Mollia*. S."

NOTE: See Introduction p. 9 for the history of the Kew sheet—F.N.H.

Polycarpon prostratum (*Forsk.*) *Asch. et Schweinf.*—FWTA. ed. 2, 1 : 131 (1954).

Exsicc.: Ada, 1784, *Isert* s.n. (**C**—1 sheet, microf. 79: I. 7).

CELASTRACEAE

Hippocratea africana (*Willd.*) *Loes. ex Engl.*—FWTA. ed. 2, 1 : 628 (1958).

Tonsella africana Willd. 1797a: 194 ("Habitat in Guinea"); Vahl 1805: 30 ("Habitat in Guinea. *Isert, Thonning*"); Schum. 1827: 20 and 1828: 40; Junghans 1961 : 339.

Exsicc.: Aflaku, 1785, *Isert* s.n. (**B**—1 sheet, No. 860, type, microf. 52: I. 5; **C**—1 sheet, type, microf. 106: III. 2); "Guinea", *Thonning* 245 (**C**—1 sheet, microf. 106: III. 3); "e Guinea" (**FI**—1 sheet; **P-JU**—1 sheet No. 12032, microf. 891: I. 2).

VERNACULAR NAME: 'Plem-tjo'.

"Grows fairly commonly".

"The white sweetish mucus which surrounds the seeds is eaten by children. Th."

Maytenus undata (*Thunb.*) *Blakelock*—FWTA. ed. 2, 1 : 624 (1958).

Celastrus lancifolius Thonning in Schum. 1827: 132 and 1828: 152; FTA. 1 : 364 (1868); Junghans 1961 : 328.

Exsicc.: Adah, *Thonning* 323 (**C**—3 sheets, type, microf. 16: II. 7, III. 1–4).

"Rare, at Adah".

CHAILLETIACEAE see DICHAPETALACEAE

CHENOPODIACEAE

Chenopodium murale *L.*—Hepper in Kew Bull. 25: 190 (1971).

C. guineense Jacq. 1788: 346–47; 1786–93a: t. 345; Schum. 1827: 156 and 1828: 176, 177 (no MS.); Junghans 1961: 329.

Exsicc.: "Guinea", *Isert* s.n. (**B**—1 sheet, No. 5338, type, microf. 363: II. 9); "Guinea", *Thonning* s.n. (**C**—1 sheet, microf. 18: II. 3).

"Here and there."
"Besides the characters cited by Willdenow *C. murale* differs from the Guinea species in having thicker more striped green stem and broader leaves, which in the young state are not downy-haired on the leaf-stalks; the common peduncles are smooth, not downy-haired; the seeds are more compressed with a distinct circular keel. S."

COMBRETACEAE

Combretum mucronatum *Schum. et Thonn.* in Schum. 1827: 184 and 1828: 204; DC. 1828: 37; Junghans 1961: 329. *C. smeathmannii* G. Don—FWTA. ed. 2, 1: 272 (1954).

Exsicc.: Aquapim, *Thonning* 158 (no type material at **C**; **G–DC**—1 sheet, 3: 20. 17, type, microf. 477: III. 4).

"In the fertile valleys at Aquapim. Flowers in Sept."

Combretum platypterum *(Welw.) Hutch. et Dalz.*—FWTA. ed. 2, 1: 274 (1954).

Combretaceae indet. in Junghans 1961: 330.

Exsicc.: Aquapim, 1786, *Isert* "153" (**C**—1 sheet, microf. 22: I. 7).

Note: It is unusual for one of Isert's specimens to bear a number, which must have been added much later. Also on the same label is the surprising note "planta parasitica".—F.N.H.

Combretum racemosum *P. Beauv.*—FTA. 2: 424 (1871); FWTA. ed. 2, 1: 272 (1954).

C. corymbosum Schumacher 1827: 185 and 1828: 205; Junghans 1961: 329.

Exsicc.: "Guinea", *Thonning* 136 (no type material at **C**; **G–DC**—1 sheet, 3: 20. 21, type, microf. 478: I. 3).

"Grows along with, and flowers at same time as the foregoing species" (i.e. *C. mucronatum*).

Conocarpus erectus *L.*—FWTA. ed. 2, 1: 280 (1954).

C. pubescens Schum. et Thonn. in Schum. 1827: 115 and 1828: 135; Junghans 1961: 329.

EXSICC.: "Guinea", *Thonning* 164 (**C**–1 sheet, type, microf. 23: III. 6); "e Guinea" (**C**—1 sheet, microf. 23: III. 4, 5).

VERNACULAR NAME: 'Mah-tjo'.

"Common in proximity of shores and salt lagoons."
"*C. erecta* much resembles our plant but differs in having smooth leaves. *C. procumbens* differs in its obovate leaves. *C. racemosa* which in its 10 stamens, obovate-oblong or top-shaped angled, singly-placed fruit and entire habit, is so different from the other species, looks as if it should constitute a distinct genus in the Tenth Class".

Uncaria inermis sensu Junghans 1962: 87, partly.

EXSICC.: "Guinea", *Thonning* "16" (**C**–1 sheet, microf. 108: I. 7).

Quisqualis indica *L.*—FWTA. ed. 2, 1: 275 (1954).

Q. obovata Schum. et Thonn. in Schum. 1827: 218 and 1828: 238: Junghans 1961: 330.

EXSICC.: Ada etc., *Thonning* 315 (**C**—1 sheet, type, microf. 90: III. 6, 7); "e Guinea" (**P–JU**—1 sheet, No. 13640, microf. 1008: III. 2).

"Grows at Ada, Töffri, Volta".
"*Q. indica* which resembles it has ovate-cordate leaves, which, especially on the lower side, are downy haired, the corolla-leaves are narrower, linear. In a second species from E. Indies the leaves are ovate-oblong acuminate, the veins beneath not downy-haired, 3–4 inches long; this is certainly different from *Q. indica*. S."

COMPOSITAE

Aspilia helianthoides (*Schum. et Thonn.*) *Oliv. et Hiern*—FWTA. ed. 2, 2: 239 (1963).

Coronocarpus helianthoides Schum. et Thonn. in Schum. 1827: 393 and 1829: 167; Junghans 1961: 330.

EXSICC.: "Guinea", *Thonning* 118 (**C**—2 sheets, types, microf. 27: I. 4, 5, only 1 sheet photographed).

"Here and there in good soil, flowers in June and July."
"Like a *Helianthus* in appearance. S."

NOTE: a second sheet has come to light since Junghans prepared his paper.—F.N.H.

Aspilia ciliata (*Schumacher*) *Wild* in Kirkia 6: 41 (1967). *A. helianthoides* (Schum. et Thonn.) Oliv. et Hiern subsp. *ciliata* (Schumacher) C. D. Adams —FWTA. ed. 2, 2: 239 (1963).

Verbesina ciliata Schumacher 1827: 391 and 1829: 165 (no MS); FTA. 3: 378 (1877); Junghans 1961: 331.

Exsicc.: "Guinea", *Thonning* s.n. (**C**—2 sheets, type, microf. 109: II. 1–4; **S**—1 sheet, type).

"Here and there."

"Approaches *Verbesina dichotoma* but in the latter the stem is dichotomously divided, the common calyx (involucre) single, the calyx leaves hairy, not hairy at margins. S".

Bidens pilosa *L.*—FWTA. ed. 2, 2: 234 (1963).

B. abortiva Schum. et Thonn. in Schum. 1827: 381; Junghans 1961: 330.

Exsicc.: Aquapim, *Thonning* 207 (no type material traced).

"A common weed in and near Aquapim."

Blumea aurita *(L. f.) DC.*—FTA. 3: 322 (1877); FWTA. ed. 2, 2: 261 (1963).

Conyza guineensis Willd. 1803: 1930 ("Habitat in Guinea"); Junghans 1961: 330.

Exsicc.: Whydah, 1785, *Isert* s.n. (**B**—1 sheet, No. 15668, type, microf. 1124: III. 1; **C**—1 sheet, type, microf. 25: III. 6, 7).

Erigeron stipulatum Schum. et Thonn. in Schum. 1827: 385 and 1829: 159.

Exsicc.: "Guinea", *Thonning* 351 (**C**—1 sheet, type, microf. 25: III. 4, 5).

Vernacular name: 'Hallasjajo'.

"In dry fields; flowers in Oct. and Nov."

"In comparing Thonning's plant with Isert's as found in latter's herbarium, it is clearly the same species although Wildenow's description differs somewhat from Thonning's. It is likewise clear that the plant is an *Erigeron*, and not a *Conyza*. S".

Conyza aegyptiaca *(L.) Ait.*—FWTA. ed. 2, 2: 254 (1963).

Erigeron exstipulatum Schum. et Thonn. in Schum. 1827: 387 and 1829: 161; Junghans 1961: 330.

Exsicc.: Asiama and Dadintam, *Thonning* 269 (**C**—1 sheet, type, microf. 40.: I. 4, 5).

"In Asiama and Dadintam. Flowers in April."

Crassocephalum rubens *(Juss. ex Jacq.) S. Moore*—FWTA. ed. 2, 2: 248 (1963).

Cacalia uniflora Schum. et Thonn. in Schum. 1827: 382 and 1829: 156; Junghans 1961: 330.

Exsicc.: "Guinea", *Thonning* 313 (**C**—3 sheets, type, microf. 13: I. 7, II. 1–5).

"Here and there, not common."

Eclipta alba (*L.*) *Hassk.* in Pl. Jav. Rar. 528 (1848); Internat. Code art. 57 (1972). *E. prostrata* (L.) L.—FWTA. ed. 2, 2 : 241 (1963).

E. punctata L.—Schum. 1827: 389 and 1829: 163; Junghans 1961 : 331.

Exsicc.: Ada, 1784, *Isert* s.n. (**C**—1 sheet, microf. 38: III. 3, 4); "Guinea", *Thonning* 317 (**C**—1 sheet, type, microf. 38: III. 1, 2; **S**—1 sheet, type); "Hb. Schum." (**C**—1 sheet, microf. 38 : II. 6, 7).

VERNACULAR NAME : 'Odiboi'

"Here and there."

Enydra fluctuans *Lour.*—FWTA. ed. 2, 2 : 242 (1963).

Caesulia radicans Willd. 1803: 1797 ("Habitat in Guinea"); Junghans 1961: 330.

Exsicc.: "Guinea", *Isert* s.n. (**B**—1 sheet, No. 15275, type, microf. 1097: I.7; **C**—perhaps microf. 110: II. 1, 2 represents an Isert sheet).

Wahlenbergia globularis Schum. et Thonn. in Schum. 1827: 387 and 1829: 161; Junghans 1961 : 331.

Exsicc.: "Guinea", *Thonning* 240 (**C**—3 sheets, type, microf. 110: I. 6, 7; II. 3–6).

"In damp shady places."
"*Caesulia radicans* Wild. is certainly the same plant but it should constitute a separate genus on account of the four-leaved common calyx, the mono-phyllously-cleft particular calyx and especially because it has bisexual flowers on the disk, female flowers at the margin. Whether Beauvois' plant belongs here is more doubtful as especially the description differs consider-ably. As the genus *Wahlenbergia* Schrader should be placed under *Campanula* as Sprengel also has assumed in his Systema, I have called it anew after the famous Swedish botanist Wahlenberg, Prof. at Uppsala. S."
"It resembles at first glance a *Spermacoce*. Th."

Erigeron spathulatum Schum. et Thonn. in Schum. 1827: 385 and 1829: 159; Junghans 1961 : 331.

Exsicc.: Aquapim, *Thonning* 228 (no type material traced).

"At Aquapim".

NOTE: Cited in FTA. 3: 308 as an unknown species which is still its status.— F.N.H.

Launaea taraxacifolia (*Willd.*) *Amin ex C. Jeffrey* in Kew Bull. 18: 474 (1966). *Lactuca taraxacifolia* (Willd.) Schum. ex Hornem. 1819: 77; FTA. 3: 451 (1877); FWTA. ed. 2, 2 : 293 (1963).

Sonchus taraxacifolius Willd. 1803: 1511 ("Habitat in Guinea"); Schum. 1827: 380 and 1829: 154; Junghans 1961 : 331.

Exsicc.: "Guinea", *Isert* s.n. (**B**—1 sheet, No. 14538, type, microf. 1039:

III. 5); "Guinea", *Thonning* 107 (**C**—3 sheets, microf. 63: III. 4–7; 64: I. 1, 2; **S**—1 sheet); "e Guinea" (**G–DC**—1 sheet 7: 138. 37, microf. 1270: III. 7).

VERNACULAR NAME: 'Abloge'.

"Grows here and there, chiefly in formerly cultivated places."

"It seems to me it should rather be referred to *Lactuca* than to *Sonchus* as the calyx is entirely like that in the first named genus and the pappus although very short is stipitate. S".

"Europeans use it as salad under the name of wild endive. It has a bitterish taste and slightly narcotic odour; the natives use the expressed sap for alleviation of pain in fresh wounds. The decoction or the leaves prepared as cabbage are used in dysentery (blood-flux). It has a similar appearance to *Lactuca scariola*. Th."

Melanthera scandens (*Schum. et Thonn.*) *Roberty*—FWTA. ed. 2, 2: 240 (1963). *Melanthera brownei* Schultz Bip.—FTA. 3: 382 (1877).

Buphthalmum scandens Schum. et Thonn. in Schum. 1827: 392 and 1829: 166; Junghans 1961: 330.

EXSICC.: "Guinea", *Thonning* 52 (**C**—2 sheets, type, microf. 12: I. 5–7; II. 1).

"Common, flowers in the fertile season of the year."

Sclerocarpus africanus *Jacq. ex Murr.*—Schum. 1827: 394 and 1829: 168; FWTA. ed. 2, 2: 235 (1963); Junghans 1961: 331.

EXSICC.: "Guinea", *Thonning* 347 (**C**—1 sheet, microf. 97: III. 3, 4); "e Guinea" (**C**—3 sheets, microf. 97: III. 5, 6; 98: I. 1–4).

"Here and there."

Spilanthes filicaulis (*Schum. et Thonn.*) *C. D. Adams*—FWTA. ed. 2, 2: 236 (1963).

Eclipta filicaulis Schum. et Thonn. in Schum. 1827: 390 and 1829: 164; Junghans 1961: 330.

EXSICC.: Aquapim, *Thonning* 227 (**C**—2 sheets, type, microf. 38: II. 2–5; **S**—1 sheet, type).

"In and near Aquapim."

"In habit and form of the flower it much resembles *Spilanthus uliginosus*; but the description shows it to be an *Eclipta*. S."

Vernonia cinerea (*L.*) *Less.*—FTA. 3: 275 (1877); FWTA. ed. 2, 2: 283 (1963).

Chrysocoma violacea Schum. et Thonn. in Schum. 1827: 384 and 1829: 158; Junghans 1961: 330.

EXSICC.: "Guinea", *Thonning* 311 (**C**—1 sheet, type, microf. 20: II. 1, 2).

"Here and there in open fields."

Vernonia colorata (*Willd.*) *Drake*—FWTA. ed. 2, 2: 277 (1963). *Vernonia senegalensis* Less.—FTA. 3: 283 (1877).

Chrysocoma amara Schum. et Thonn. in Schum. 1827: 383 and 1829: 157; Junghans 1961: 330.

EXSICC.: Quitta & L. Augna, *Thonning* 131 (**C**—1 sheet, type, microf. 20: I. 6, 7).

VERNACULAR NAME: 'Tah-tjo'.

"Here and there, although not common; at Quitta and on the islands in Augna Lake. Flowers in Nov."

"The natives employ this plant for various medicinal uses; for old bone injuries a decoction of the leaves is used for bathing; the soft beaten and moistened bark of the root is applied to the wound itself; where the blood flux is not too serious a decoction of the leaves with a little grains-of-Paradise [*Aframomum*] is used for drinking; in rheumatic pains the leaves expressed in cold water for a bath are used, and thereafter the limb is smeared with the finely grated root; but some fetish is connected with this last operation. The root but especially the leaves have a rather clean but very bitter taste. I have seen Europeans employ it for a bitter essence which has been very good. Th."

CONNARACEAE

Agelaea obliqua (*P. Beauv.*) *Baill.*—FWTA. ed. 2, 1: 745 (1958).

Connaraceae indet. in Junghans 1961: 332.

EXSICC.: Whydah, 1785, *Isert* s.n. (**C**—1 sheet, microf. 3. II. 7).

Byrsocarpus coccineus *Schum. et Thonn.* in Schum. 1827: 226 and 1828: 246; FWTA. ed. 2, 1: 741 (1958); Junghans 1961: 331, 332.

EXSICC.: "Guinea", *Thonning* 19 (**C**—10 sheets, type, microf. 12: II. 6, 7; III. 1–7; 13: I. 1).

VERNACULAR NAME: 'Plöm-tjo' and fruit, 'Sjo-tami'.

"Common".

"The bark is scraped off the fresh root, or failing that, off the dried one, is beaten into a soft pulp and applied to an old leg injury; also the wound is bathed with a decoction. William Parker has assured me that he has seen a native take the leaves of this plant along with a few grains-of-Paradise [*Aframomum*], chew them and apply them to the bite of a poisonous snake, and that he afterwards found him quite well; but he did not know of what species the snake was. Other natives assured me that it was not the healing power of the plant which had been effective here, but that it was a fetish cure which could help him, but no one else, as the plant was his fetish. Th."

B. puniceus Schum. et Thonn. in Schum. 1827: 227 and 1828: 247.

EXSICC.: "Guinea", 1784, *Isert* s.n. (**C**—1 sheet, microf. 12: II. 5); "Guinea", *Thonning* 14 (**C**—2 sheet, types, microf. 12: II. 2–4).

VERNACULAR NAME: 'Plöm-tjo' and fruit, 'Sio-tami'.

"Grows with the preceding one". [i.e. *B. coccineus*]

"Use as in the preceding species. Th."

"Although the stated difference between the length relations of the male and female parts is constant, it nevertheless requires further investigation, as to whether the differences do not arise from the fact that the one kind of plant is female with incomplete stamens, the other a male plant with incomplete pistil; in that case the genus would belong to Dioecia and approaches *Zanthoxylon* or may perhaps be united with it. S."

Cnestis ferruginea *DC.*—FWTA. ed. 2, 1: 743 (1958); Junghans 1961: 332.

EXSICC.: Fida, 1784, *Isert* s.n. (C—1 sheet, microf. 21: III. 6.); "Guinea", *Thonning* s.n. (C—2 sheets, microf. 21: II. 7, III. 1–3); "e Guinea" (C—1 sheet, microf. 21: III. 4, 5; P–JU—1 sheet, No. 15977A, microf. 1160: I. 4).

Connarus africanus *Lam.*—FWTA. ed. 2, 1: 748 (1958).

Connaraceae indet. in Junghans 1961: 332.

EXSICC.: "Guinea, Hb. Héritier" (C—1 sheet, microf. 23: III. 3).

Connarus thonningii (*DC.*) *Schellenb.*—FWTA. ed. 2, 1: 748 (1958).

Omphalobium thonningii DC. 1825a: 86; Junghans 1961: 332.

EXSICC.: "Guinea", *Thonning* 339? (G–DC—1 sheet, 2: 86. 7, type, microf. 294: III. 7).

Connarus floribundus Schum. et Thonn. in Schum. 1827: 299 and 1829: 73.

EXSICC.: Töffri, *Thonning* 339 (C—3 sheets, types, microf. 23: I. 5–7, II. 1–3); "e Guinea" (C—1 sheet, microf. 23: II. 4, 5, the leaves on the left do not belong to this specimen.—F.N.H.)

"From Töffri."

"Agrees with *C. nemorosus* Vahl (Mpt.) but the leaves in this last are tridigitate or pinnate, the leaflets ovate pointed; the flowers smaller, petals shorter, linear-cuneate; receptacle rather stiff haired not silky haired. *C. pinnatus* differs with tridigitate and pinnate leaves, and with 2 bristles at base of petals. S."

NOTE: *Connarus nemorosus* mentioned above is the Asiatic *C. monocarpus* L. based on Indian material collected by König (probably IDC 2203 microf. 23. II. 6, 7). One sheet of Rottler's Indian material also appears in microf. 23: III. 1–2).—F.N.H.

CONVOLVULACEAE

Aniseia martinicensis (*Jacq.*) *Choisy*—FWTA. ed. 2, 2: 343 (1963).

Convolvulaceae indet. in Junghans 1961: 334.

EXSICC.: Whydah, 1785, *Isert* s.n. (C—1 sheet, microf. 60: II. 3).

Evolvulus alsinoides (*L.*) *L.*—FTA. 4, 2: 67 (1905); FWTA. ed. 2, 2: 339 (1963).

E. azureus Vahl ex Thonn. in Schum. 1827: 166 and 1828: 186; Junghans 1961: 333.

Exsicc.: Ada, 1784, *Isert* s.n. (**C**—1 sheet, microf. 41: III. 5; **M**—1 sheet; **S**—1 sheet); "Guinea", *Thonning* 162 (**C**—5 sheets, type, microf. 41: III. 6, 7; 42: I. 1–3; **S**—1 sheet, type; **LE**—1 sheet, type).

"Here and there, usually in dry places".
"*E. hirsutus* differs in having ovate leaves hairy on both sides, and floral leaves in the middle of the flower-stalks. S."

Hewittia sublobata (*L. f.*) *Kuntze*—FWTA. ed. 2, 2: 342 (1963).

Convolvulus involucratus Willd. 1798a: 845 ("Habitat in Guinea"); Schum. 1827: 93 and 1828: 113; Junghans 1961: 333.

Exsicc.: Whydah, 1785, *Isert* s.n. (**B**—1 sheet, No. 3630, type, microf. 250: III. 8; **C**—1 sheet, type, microf. 24: III. 3); Aquapim, *Thonning* 208 (**C**—3 sheets, microf. 24: II. 6, 7; III. 2); "e Guinea" (**C**—1 sheet, microf. 24: III. 1; **P–JU**—1 sheet, No. 6764, microf. 494: III. 1).

"Rarely in Aquapim."
"In the specimen lying in Isert's herbarium all the leaves are cordate—oblong and obtuse, and the flower-stalks not so long as the leaves. It is therefore rather uncertain whether it may be merely the upper portion of the plant or another species."

Ipomoea cairica (*L.*) *Sweet*—FWTA. ed. 2, 2: 351 (1963).

Convolvulus cairicus L.—Schum. 1827: 96 and 1828: 116; Junghans 1961: 332.

Exsicc.: Ada, 1784, *Isert* s.n. (**C**—1 sheet, microf. 25: I. 3); Whydah, 1785, *Isert* s.n. (**C**—1 sheet, microf. 25: I. 4); R. Volta & Sakumo, *Thonning* 253 (**C**—4 sheets, microf. 25: I. 1, 5–7); "e Guinea" (**C**—1 sheet, microf. 25: I. 2).

"Chiefly at the large rivers e.g. Volta & Sakumo, it winds high up on the bushes."

Ipomoea coptica (*L.*) *Roth ex Roem. et Schult.*—FWTA. ed. 2, 2: 350 (1963).
Convolvulus pinnatifidus Schumacher ex Hornem. 1819: 16.

I. dissecta Willd. 1794: 5, Pl. 2, f. 3 ("Habitat in Guinea. *Isert*"); 1798a: 880; FTA. 4, 2: 176 (1905); Junghans 1961: 334.

Exsicc.: Ada, 1784, *Isert* s.n. (**B**—1 sheet, No. 3746, type, microf. 258: II. 3; **C**—1 sheet, type, microf. 60: III. 6).

C. thonningii Schum. et Thonn. in Schum. 1827: 98 and 1828: 118.

Exsicc.: "Guinea", *Thonning* 5 (**C**—4 sheets, type, microf. 60: III. 7; 61: I. 1–3); "e Guinea" (**FI**—1 sheet; **P–JU**—1 sheet, No. 6850 B, microf. 501: I. 1).

"Not very common; flowers in May."

"The specimen which is in Isert's herbarium and is found at Ada, is in all parts larger, flower- and leaf-stalks nearly twice as long, one-flowered except one which bears 2 flowers. Very nearly related to our plant are *C. copticus* and *C. laciniatus* Lamarck but in the former the leaf's middle lobe is obovate and the lateral lobes less divided, the calyx-segments more prickly and the floral leaves on the single flower-stalks palmate; the latter is distinguished by the whole plant having stiff hairs. S".

Ipomoea sepiaria *Roxb.*—Verdcourt in FTEA. Convolv. 117 (1963). *I. hellebarda* Schweinf. ex Hiern—FTA. 4, 2: 170 (1905); FWTA. ed. 2, 2: 349 (1963).

Convolvulus diversifolius Schum. et Thonn. in Schum. 1827: 94 and 1828: 114; Junghans 1961: 333.

EXSICC.: "Guinea", *Thonning* 181 (**C**—3 sheets, type, microf. 25: II. 4–7); "e Guinea" (**P–JU**—1 sheet No. 6766, type, microf. 494: III. 3).

"Here and there but not common. Flowers in June."

Ipomoea heterotricha *F. Didr.* 1854: 220 ("e Congo. C. Smith. e Guinea mis. Mortensen"); FWTA. ed. 2, 2: 347 (1963); Junghans 1961: 334.

EXSICC.: "*C. Smith and Mortensen*" s.n. (**C**—1 sheet, type, 61: I. 4); "Guinea", *Mortenson* s.n. (**C**—1 sheet, type, microf. 61: I. 5); "Congo", *C. Smith* s.n. (**C**—1 sheet, type, microf. 61: I. 7); "Guinea", *Thonning* s.n. (**C**—1 sheet, microf. 61: I. 6).

Ipomoea involucrata *P. Beauv.*—FTA. 4, 2: 150 (1905); FWTA. ed. 2, 2: 374 (1963).

Convolvulus perfoliatus Schum. et Thonn. in Schum. 1827: 89 and 1828: 109; Junghans 1961: 333.

EXSICC.: Whydah, 1785, *Isert* s.n. (**C**—1 sheet, microf. 24: I. 2); Aquapim, *Thonning* 209 (**C**—2 sheets, type, microf. 24: I. 1, 3).

"Occurs in Aquapim, but rare".

Ipomoea mauritiana *Jacq.*—FWTA. ed. 2, 2: 351 (1963).

Convolvulus paniculatus L.—Schum. 1827: 94 and 1828: 114; Junghans 1961: 333.

EXSICC.: "Guinea," *Thonning* 21 (**C**—2 sheets, microf. 24: III. 4, 5); "e Guinea" (**P–JU**—1 sheet, No. 6845, microf. 500: II. 7).

VERNACULAR NAME: 'Löloa-pang'.

"Common."

"I dare not definitely decide whether our plant is the same as that of Linnaeus and the other authors; but it is certain that the authors' β is a quite different plant, as it has equally long and unbent flower-stalks, 3-lobed leaves with broad-lanceolate lobes and with 2 glands at the base, and a rhomboid gland in each sinus also the corollas are much smaller. For dropsy both under the skin and in the abdomen the natives boil the root with some grains-of-Paradise (Guinea pepper) [*Aframomum*]; for drinking and for external application the root is ground fine with some grains-of-Paradise and a little water and rubbed over the whole body. For gonorrhea virulenta the root is finely broken up and placed in pytto or palm-wine which is put aside to become sour and then is used for promoting urine. In absence of string the natives use the stems for binding fire wood and such like."

Ipomoea ochracea (*Lindl.*) *G. Don*—FTA. 4, 2: 166 (1905); FWTA. ed. 2, 2: 349 (1963).

Convolvulus trichocalys Schum. et Thonn. in Schum. 1827: 91 and 1828: 111; Junghans 1961: 333.

Exsicc.: "Guinea", 1786, *Isert* s.n. (**C**—1 sheet, microf. 24: II. 1); "Guinea," *Thonning* 6 (**C**—2 sheets, type, microf. 24: I. 6, 7); "e Guinea" (**P–JU**—1 sheet, No. 6760, type, microf. 494: I. 5).

"Here and there".
"*C. gemellus* differs from our plant in that the leaves are cordate, less pointed, downy-hairy on both sides, the flower-stalks not longer than the leaf-stalks, 2-flowered and like the stem weakly downy-haired."

Convolvulaceae indet. in Junghans 1961: 334.

Exsicc.: "Guinea," *Mortensen* s.n. (**C**—1 sheet, microf. 60: II. 2).

Ipomoea pes-caprae (*L.*) *R. Br.* subsp. **brasiliensis** (*L.*) *Oostr.*—FWTA. ed. 2, 2: 347 (1963). *I. biloba* Forsk.—FTA. 4, 2: 172 (1905).

Convolvulus rotundifolius Schum. et Thonn. in Schum. 1827: 102 and 1828: 122; Junghans 1961: 333.

Exsicc.: Töffri, *Thonning* 11 (**C**—2 sheets, type, microf. 24: I. 4, 5).

Vernacular name: 'Amba-pang'.

"Common on shore-margins in the loose sand. The sub-species is found at Töffri in places which are flooded by the river Volta."
"*C. pes caprae* is distinguished from it in having two-three flowered peduncles and two-lobed leaves, *C. brasiliensis* in having leaf-stalks which are 4 inches long and longer than the peduncles. S".

Ipomoea stolonifera (*Cyrill.*) *J. F. Gmel.*—FTA. 4, 2: 171 (1905); FWTA. ed. 2, 2: 350 (1963).

Convolvulus incurvus Schum. et Thonn. in Schum. 1827: 99 and 1828: 119; Junghans 1961: 333.

Exsicc.: "Guinea" shore, *Thonning* 230 (**C**—4 sheets, types, microf. 24: II. 2–5; **P–JU**—1 sheet, No. 6877, microf. 503: III. 4).

Vernacular name: 'Vula-fyé'

"Common on shore margins in the loose sand".
"The leaves are used by the natives as cabbage. Th."
"*C. acetosaefolius* differs from our plant in having broader leaves, fewer nearly indistinct lobes, flower stalks which are double as long as the leaves and

alternate floral leaves. *C. emarginatus* likewise in having broader leaves, which are sinuate at the base, flower-stalks which are shorter than the leaves and the outer lobes of the calyx larger. The figure in Plum. Am. tab. 105 does not fit *C. acetosaefolius* in Vahl's herbarium, and Burman's figure Ind. tab. 21, fig. 2 does not at all fit *C. emarginatus* in the same herbarium. S".

Jacquemontia ovalifolia *(Vahl)* *Hall.* f.—FTA. 4, 2: 87 (1905); FWTA. ed. 2, 2: 340 (1963).

Convolvulus coeruleus Schum. et Thonn. in Schum. 1827: 101 and 1828: 121; Junghans 1961: 332.

Exsicc.: "Guinea," *Thonning* 62 (**C**—3 sheets, type, microf. 25: II. 1–3); "e Guinea" (**FI**—1 sheet; **P–JU**—1 sheet, No. 6738, microf. 492: III. 2).

Vernacular name: 'Klovaké' (Accra dialect).

"Fairly common on the shore".

"*C. ovalifolius* Vahl Eclog. 2p. 16. resembles this plant so much that one would take it for the same species, it differs only in the erect stem, the flower-stalks which are longer than the leaf-stalks, one–seven-flowered, and have 2–3 or 4 floral leaves at the base of the pedicels. It is therefore uncertain whether our plant is merely a subspecies or a species on its own. S."

"The natives regard the leaves as a good cabbage. The fetish of the Ussu tribe eats this cabbage wherefore none of this tribe dare taste it and the infringement of this command is regarded as a great crime".

Jacquemontia tamnifolia *(L.)* *Griseb.*—FWTA. ed. 2, 2: 340 (1963).

Convolvulus guineensis Schumacher 1827: 90 and 1828: 110; Junghans 1961: 333.

Exsicc.: "Guinea" shore, *Thonning* 190 (**C**—3 sheets, type, microf. 25: III. 1–3); "e Guinea" (**FI**—1 sheet; **P–JU**—1 sheet No. 6803, type, microf. 497: II. 7).

"On the shore, but rare. Flowers in May, June and July."

"Perhaps it is the same plant which is in Vahl's herbarium under the name of *C. ciliatus* sent by Rohr from Cayenne."

Merremia aegyptiaca *(L.)* *Urban*—FWTA. ed. 2, 2: 342 (1963).

Convolvulus pentaphyllus L.—Schum. 1827: 97 and 1828: 117; Junghans 1961: 333.

Exsicc.: Ada, 1784, *Isert* s.n. (**C**—1 sheet, microf. 24: III. 7); "Guinea," *Thonning* 2 (**C**—1 sheet, microf. 24: III. 6).

"Here and there near the towns among *Cactus tuna*."

"A specimen of this plant is also in Isert's herbarium."

Merremia tridentata *(L.)* *Hall.* f. subsp. **angustifolia** *(Jacq.)* *Oostr.*—FWTA. ed. 2, 2: 341 (1963). *M. angustifolia* (Jacq.) Hall. f.—FTA. 4, 2: 111 (1905).

Ipomoea angustifolia Jacq. 1788: 367–368 ("Ex Guinea est"); 1786–93a: t. 317; Junghans 1961: 334.

Exsicc.: Guinea, *Isert* s.n. (**B**—2 sheets, No. 3638, type, microf. 251: II. 3, 4).

Convolvulus filicaulis Vahl 1794: 24 ("In Guinea invenit Dn. Isert qui Semina Dn. von Rohr dedit unde plantam habuit Dn. Lund"); Willd. 1798a: 848; Schum. 1827: 92 and 1828: 112.

Exsicc.: "Guinea", *Isert* s.n. (**C**—2 sheets, type, microf 60: II. 7; III. 1); "Guinea", *Thonning* 87 (**C**—6 sheets, microf. 60: II. 5, 6; III. 2–5); "e Guinea" (**C**—1 sheet, microf. 60: II. 4; **FI**—1 sheet; **P–JU**—1 sheet, No. 6767, microf. 494: III. 4).

"Common in shore-landscapes; it twines up in the tall grass and is found flowering now and then."

"Very closely related to *Evolvulus tridentatus* (*Convolvulus tridentatus*) Vahl, l.c. Perhaps *Ipomaea angustifolia* Jacq. collect. 2p. 367 belongs here. S."

NOTE: In the Willdenow Herbarium Index the Isert specimens appear under *C. filicaulis*. No doubt they are all from the same gethering.—F.N.H.

CRASSULACEAE

Kalanchoe crenata (*Andr.*) *Haw.*—FWTA. ed. 2, 1: 118 (1954).

Verea crenata Andr.—Schum. 1827: 199 and 1828: 219 (no MS); Junghans 1961: 334.

Exsicc.: *Thonning* s.n.? (no material traced).

CUCURBITACEAE

Cucumis melo *L.* var. **agrestis** *Naud.*—FTA. 2: 546 (1871); FWTA. ed. 2, 1: 213 (1954).

Exsicc.: Ada, 1784, *Isert* s.n. (**C**—2 sheets, microf. IDC set 2204, 61: II. 1–4).

NOTE: The sheets at Copenhagen were included in the microfiche set of the Type Herbarium of Museum Botanicum Hauniense; the IDC label gives the wrong reference which is corrected here.—F.N.H.

C. arenarius Schum. et Thonn. in Schum. 1827: 426 and 1829: 200; Junghans 1961: 335.

Exsicc.: "Guinea", *Thonning* 151 (**C**—1 sheet, type, microf. nil); "e Guinea" (**C**—1 sheet, microf. nil; **P–JU**—1 sheet No. 16622, microf. 1202: III. 5).

VERNACULAR NAME: 'Nanni-adumatre'; Danish: 'Nannis vandmelon'.

"Grows in dry sandy soil."

Kedrostis foetidissima (*Jacq.*) *Cogn.*—FWTA. ed. 2, 1: 210 (1954).
Rhynchocarpa foetida Schrad.—FTA. 2: 564 (1871).

Trichosanthes foetidissima Jacq. 1788: 341-2 ("In Guinea sponte crescit");
 1786–93b: t. 624; Willd. 1805: 599; Junghans 1961: 335.

EXSICC.: "Guinea", *Isert* s.n. (**B**—2 sheets, No. 18018, types, microf. 1301:
III. 2, 3).

Bryonia foetidissima Schum. et Thonn. in Schum. 1827: 428 and 1829: 202
 (MS No. 337).

EXSICC.: Fida, 1784, *Isert* s.n. (**C**—1 sheet, type, microf. nil).

VERNACULAR NAME: 'Sia-pang'.

"Here and there".
"The whole plant is extremely fetid almost like rotting cabbage. It is
boiled in water for bathing against tenesmus (bowel straining). Th."

Lagenaria breviflora (*Benth.*) *G. Roberty*—Jeffrey in J.W. Afr. Sci. Assoc.
9: 90 (1964). *Adenopus breviflorus* Benth.—FWTA. ed. 2, 1: 206 (1954).

EXSICC.: Whydah, *Isert* s.n. (**C**—1 sheet, microf. nil).

Lagenaria siceraria (*Molina*) *Standley*—FWTA. ed. 2, 1: 206 (1954).

EXSICC.: Ada, 1784, *Isert* s.n. (**C**—1 sheet, as "Cucurbita maxima n.d.,"
microf. nil).

Cucurbita idololatrica Willd. 1805: 607 ("Habitat in Guinea"); Junghans 1961:
 335.

EXSICC.: Ada, 1784, *Isert* s.n. (**B**—1 sheet, No. 18030, type, microf. 1302:
II. 7; **C**—1 sheet, type, microf. nil).

Luffa aegyptiaca *Mill.*—FTA. 2: 530 (1871); FWTA. ed. 2, 1: 207 (1954).

L. scabra Schum. et Thonn. in Schum. 1827: 405 and 1829: 179; Junghans
 1961: 335.

EXSICC.: "Guinea" coast, *Thonning* 278 (**C**—1 sheet, type, microf. nil).

"In the native communities of the coast amongst *Cactus tuna*."
"I am doubtful if this species ought not to constitute a separate genus. S."

Momordica charantia L.—FTA. 2: 537 (1871); FWTA. ed. 2, 1: 212
(1954).

M. anthelmintica Schum. et Thonn. in Schum. 1827: 423 and 1829: 197;
 Junghans 1961: 335.

EXSICC.: "Guinea", *Thonning* 3 (no type material traced.)

VERNACULAR NAME: 'Jan-j'na'.

"Grows here and there, chiefly in the proximity of towns."
"It flowers almost continuously: the natives' children eat the fruit; its
soft valves are watery, but the mucus which enwraps the seeds, is not very
sweet. The leaves have an unpleasant disgusting odour. The natives use it
against worms (*Lumbricus*) in the following way: 1 to 2 good handfuls of the

fresh plant are squeezed with about ½ pot of water, to which is added the juice of 4 lemons, a little stone weighing 6–8 half-ounces is made red hot and thrown therein. When the mixture has become cold again, it is drunk, the result is either vomiting or bowel-action or both whereby the worms are expelled. The same drink is used for a re-opening in costiveness. Besides these uses it is one of the most important of the large number of fetish-plants. Most fetish ceremonies generally finish with washing the body with water in which consecrated leaves are laid, this plant is used on many of these occasions; when a native on his journeys finds it, he gladly slings a piece round his neck with the idea that it will assuredly preserve him from unfortunate happenings. Th."

"Much resembles *M. charantia* but in the latter the leaves are palmate not palmately-pedate (palmato-pedata). The leaves of the flowers are situated below the middle of the flower-stalk. *M. balsamina* has five-lobed palmate smooth leaves which as in the allied species e.g. *M. muricata* have pointed teeth and the lobes do not taper to the base. S."

Momordica foetida *Schumacher* 1827: 426 and 1829: 200 (no MS); FWTA. ed. 2, 1 : 212 (1954); Junghans 1961 : 335.

Exsicc.: "Guinea", *Thonning* 85 (**C**—1 sheet, type, microf. IDC set 2204, 125: III. 1–2; **LE**—1 sheet, type).

"Here and there."
"I am uncertain if it belongs to this genus, as a description of the inner parts of the flower is wanting. S."

Rhaphidiocystis chrysocoma (*Schumacher*) *C. Jeffrey* in Kew Bull. 15: 60 (1962); Keraudren & Jeffrey in Bull. Jard. Bot. Nat. Belg. 37: 325 (1967). *Cucumis ficifolius* A. Rich.—FWTA. ed. 2, 1: 213 (1954).

C. chrysocomus Schumacher 1827: 427 and 1829: 201 (no MS); Junghans 1961 : 335.

Exsicc.: "Guinea", *Thonning* s.n. (**C**—1 sheet, type, microf. IDC set 2204, 61: II. 5, 6; **M**—1 sheet, type).

"Rare."
"I am uncertain if it belongs to this genus or another or constitutes a separate one."

Note: The sheet at Copenhagen was included in the microfiche set of the Type Herbarium of Museum Botanicum Hauniense; an error on the IDC label is corrected here.—F.N.H.

Zehneria capillacea (*Schum. et Thonn.*) *C. Jeffrey* in Kew Bull. 15: 366 (1962).

Melothria capillacea (Schum. et Thonn.) Cogn.—FWTA. ed. 2, 1 : 209 (1954).

Bryonia capillacea Schum. et Thonn. in Schum. 1827: 430 and 1829: 204 (MS No. 318); Junghans 1961 : 334.

Exsicc.: Ada, 1784, *Isert* s.n. (**C**—1 sheet, microf. nil); "Herb. Schum." probably *Thonning* 318 (**LE**—1 sheet, type).

"At Ada."

Zehneria hallii *C. Jeffrey* in J.W. Afr. Sci. Assoc. 9: 93 (1964).

Melothria deltoidea (Schum. et Thonn.) Benth.—FTA. 2: 563 (1871); FWTA. ed. 2, 1: 209 (1954).

Bryonia deltoidea Schum. et Thonn. in Schum. 1827: 429 and 1829: 203.

Exsicc.: "Guinea", *Thonning* 153 (**C**—1 sheet, type, microf. nil; **LE**—1 sheet, type); "e Guinea" (**P–JU**—1 sheet No. 16606, microf. 1201: II. 7).

"Here and there. Flowers in July."

Cucurbitaceae: 5 Isert specimens cited by Junghans 1961: 335 as indeterminate, were not traced in 1971—F.N.H.

DICHAPETALACEAE

Dichapetalum guineense (*DC.*) *Keay*—FWTA. ed. 2, 1: 436 (1958). *Ceanothus guineensis* DC. 1825a: 30; Junghans 1961: 329. Type as below.

Rhamnus paniculatus Thonning in Schum. 1827: 131 and 1828: 151; FTA. 1: 341 (1868).

Exsicc.: "Guinea", *Isert* s.n. (**C**—1 sheet, microf. 92: I. 3, 4); "Guinea" shore, *Thonning* 289 (**C**—3 sheets, types, microf. 91: III. 6, 7; 92: I. 1, 2; **G–DC**—1 sheet, 2: 30. 10, type, microf. 280: II. 3).

Vernacular name: 'Ototrómi'.

"Common in shore coastal areas."

DILLENIACEAE

Tetracera alnifolia *Willd.* 1800a: 1243 ("Habitat in Guinea"); FWTA. ed. 2, 1: 180 (1954); Junghans 1961: 335.

Exsicc.: Whydah, 1785, *Isert* s.n. (**B**—2 sheets, No. 10353, types, microf. 722: III. 5, 6; **C**—1 sheet, type, microf. nil).

EBENACEAE

Diospyros ferrea (*Willd.*) *Bakh.*—FWTA. ed. 2, 2: 11 (1963). *Maba buxifolia* Pers.—FTA. 3: 515 (1877).

Ferreola guineensis Schum. et Thonn. in Schum. 1827: 448 and 1829: 222; Junghans 1961: 335.

Exsicc.: Quitta, *Thonning* 343 (**C**—3 sheets, type, microf. 42: I. 6, 7; II. 1–3).

"At Quitta."

"Wonder if Persoon was right in uniting *Maba* and *Ferreola*? In specimens of *Maba elliptica* of Forster's collecting which are in Vahl's herbarium, the calyx of the female flower is 5-cleft, nearly 5-partite, the lobes lanceolate, pointed, smooth. S."

Diospyros tricolor *(Schum. et Thonn.)* *Hiern*—FTA. 3: 521 (1877); FWTA. ed. 2, 2: 12 (1963).

Noltia tricolor Schum. et Thonn. in Schum. 1827: 189 and 1828: 209; Junghans 1961: 336.

Exsicc.: "Guinea", *Isert* "332" or "689" (**C**—1 sheet microf. 42: I. 4, 5); "Guinea", shore, *Thonning* 252 (**C**—5 sheets, type, microf. 70: I. 6, 7; II. 1–7; III. 1).

Vernacular name: 'Aumbae'.

"Common in proximity to shore-margins."
"I have named the genus after the excellent botanist Professor Nolte, known for his treatise on *Stratiotes* and *Sagittaria* and his Novitiae Florae Holsaticae. S."

Note: Junghans placed the Isert sheet, with a query, under *Ferreola guineensis*—F.N.H.

ERYTHROXYLACEAE

Erythroxylum emarginatum *Thonning* in Schum. 1827: 224 and 1828: 244; FWTA. ed. 2, 1: 356 (1958); Junghans 1961: 336.

Exsicc.: Christiansborg, 1783, *Isert* s.n. (**C**—1 sheet, microf. 40: II. 1); "Guinea", *Thonning* 290 (**C**—3 sheets, type, microf. 40: II. 2–4); "e Guinea" (**FI**—1 sheet, No. 28647; **P–JU**—1 sheet, No. 11455, type, microf. 849: I. 5).

Vernacular name: 'Sio-tahmi'.

EUPHORBIACEAE

Acalypha ciliata *Forsk.*—Willd. 1805: 522; FTA. 6, 1: 901 (1912); FWTA. ed. 2, 1: 410 (1958).

Exsicc.: "Guinea", *Isert* s.n. (**B**—1 sheet, No. 17807, microf. 1289: II. 3).

A. fimbriata Schum. et Thonn. in Schum. 1827: 409 and 1829: 183; Junghans 1961: 336.

Exsicc.: Aquapim, 1786, *Isert* s.n. (**C**—1 sheet, microf. 2: I. 1–2); "Guinea", *Thonning* 275 (**C**—4 sheets, type, microf. 1: II. 5–7; III. 1–5); "e Guinea" (**C**—1 sheet, microf. 1: III. 6–7; **P–JU**—1 sheet, No. 16537, microf. 1197: I. 1).

"Here and there."
"In habit much like *A. ciliata* Vahl, Symb. Bot. I p. 77 tab. 20, with which Wildenow (sic) has wrongly united our plant (Sp. pl. 4 p. 522); in *A.*

ciliata the female flowers are more clustered and the spikes much shorter; the leaves near the flowers have longer and hairy-margined lobes and the stems are more deeply furrowed. S."

Alchornea cordifolia (*Schum. et Thonn.*) *Müll. Arg.*—FWTA. ed. 2, 1: 403 (1958). *A. cordata* Benth.—FTA. 6, 1: 915 (1912).

Schousboea cordifolia Schum. et Thonn. in Schum. 1827: 449 and 1829: 223; Junghans 1961: 338.

EXSICC.: Whydah, 1785, *Isert* s.n. (**C**—1 sheet, microf. 96: I. 4); "Guinea", *Thonning* 316 (**C**—3 sheets, type, microf. 96: I. 1–3); "e Guinea" (**C**—1 sheet, microf. 96: I. 5–6; **P–JU**—1 sheet, No. 16524a, microf. 1196: I. 5).

"Here and there."

"As Persoon and Decandolle following Aublet have called Wildenow's Schousboea: Cacoutia; I have renamed it after K.A. Schousboe, Legation Councillor, Danish General-Consul in Morocco, Knight of Dannebrog. S."

Bridelia ferruginea Benth.—FWTA. ed. 2, 1: 370 (1958).

Euphorbiaceae indet. in Junghans 1961: 338.

EXSICC.: Whydah, 1785, *Isert* s.n. (**C**—1 sheet, microf. 21: II. 6); "e Guinea" [prob. Isert] (**C**—1 sheet, microf. 11: II. 1, 2); "Herb. Rottboll" [prob. Isert] (**C**—1 sheet, microf. 21: II. 3, 4); "Guinea", *Thonning* s.n. (**C**—2 sheets, microf. 11: II. 3, 4; 21: II. 5).

Croton lobatus *L.*—FTA. 6, 1: 750 (1912); FWTA. ed. 2, 1: 394 (1958). *Schradera scandens* Willd. 1797b: 3, pl. 2 ("Habitat in Guinea. *Isert*"); 1798b: 132, pl. 7; Junghans 1961: 338.

C. trilobatum Forsk.—Willd. 1805: 556 as to plant from Guinea; Schum. 1827: 411 and 1829: 185, 186 as to plant from Guinea.

EXSICC.: "Guinea", *Isert* s.n. (**B**—1 sheet, No. 17910, microf. 1295: I. 7); "Guinea", *Thonning* 116 (**C**—1 sheet, microf. 28: III. 3, 4).

"Here and there, not rare. Flowers in fertile season of year."

"*C. lobatum* differs from our plant in that the hairs on the leaves are longer, placed singly rarely two together and never stellate; in that the twigs and upper side of the leaf-stalks are beset with long hairs and not with short and felt-like hairs; in that the calyx-leaves in the female flowers are twice as long and not whitish-grey. Persoon was therefore hardly right in uniting these plants. In Vahl's herbarium there is a plant under the name of *C. trilobatum* found by Forskaal in Arabia which is entirely our plant, except that it is rather smaller, like most plants from there. S."

Elaeophorbia drupifera (*Thonning*) *Stapf*—FWTA. ed. 2, 1: 423 (1958).

Euphorbia drupifera Thonning in Schum. 1827: 250 and 1829: 24; Junghans 1961: 336.

EXSICC.: *Thonning* 266 (no type material traced).

VERNACULAR NAME: 'Tenjo-tjo'.

"The whole plant is abundantly rich in a milk-white sap, which is so sharp that the least drop almost inevitably deprives one of sight. It coagulates immediately both in the air, in water and in strong spirit, to a whitish resin which is opaque, porous, floats on water, burns with a reddish flame and has no odour. Natives from the interior of the country use a decoction of it for washing exulcerationes gingivae. In some places it is used for casting in the water in order to stupefy fish. Small fishes die. The stem is utilised very occasionally for making native drums. Th."

Euphorbia glaucophylla *Poir.*—FTA. 6, 1 : 499 (1912); FWTA. ed. 2, 1 : 419 (1958).

E. trinervia Schum. et Thonn. in Schum. 1827: 253 and 1829: 27; Junghans 1961: 337.

EXSICC.: "Guinea" shore, *Thonning* 32 (**C**—2 sheets, types, microf. 41 : III. 1–4).

"Common on shore-margins."
"Closely allied to *E. glaucophylla* but the latter has a dichotomously divided stem, tripartite umbels and nearly entire-margined leaves. Perhaps it comes even nearer to *E. glabrata* except that the flowers in the latter occur in heads in the leaf-axils. S."

Euphorbia lateriflora Schum. et Thonn. in Schum. 1827: 252 and 1829: 26; FWTA. ed. 2, 1 : 422 (1958); Junghans 1961 : 336.

EXSICC.: "Guinea", *Thonning* 298 (**C**—1 sheet, type, microf. 41 : I. 1, 2); "e Guinea?" (**C**—1 sheet, inadequate, microf. 41 : I. 3, 4).

"In thickets."
"Different from *E. mauritanica* which comes nearest to it; this latter has branches and twigs provided with adjacent raised rough transverse lines which are the remains of the fallen leaves; and not with distant scattered nodes; the leaves are narrower, the umbels are terminal and bear 5–8 flowers; not laterally with 3–4 flower-stalks. *E. piscatoria* has a rigid stem, narrower leaves, terminal umbels and many flowers. S."

Euphorbia prostrata *Ait.*—FWTA. ed. 2, 2 : 421 (1958).

E. chamaesyce sensu Schum. 1827: 254 and 1829: 28 (no MS.); Junghans 1961 : 366, non L.

EXSICC.: "Guinea", *Thonning* s.n. (**C**—2 sheets, types, microf. 40 : III. 4–7); "e Guinea" (**P–JU**—1 sheet, No. 16392, microf. 1185 : III. 6).

"In dry places."
"Most closely allied to *E. granulata* but the latter is quite downy haired, the branches are reddish, the leaves entire-margined and smaller. S."

Euphorbia purpurascens *Schum. et Thonn.* in Schum. 1827: 252 and 1829: 26; FWTA. ed. 2, 1 : 422 (1958); Junghans 1961 : 336.

Exsicc.: Ada, 1785, *Isert* s.n. (**C**—1 sheet, microf. 41: II. 6, 7); "Guinea", *Thonning* 237 (**C**—4 sheets, types, microf. 41: I. 5–7, II. 1–5); "e Guinea" (**FI**—1 sheet; **P–JU**—1 sheet, No. 16390, microf. 1185: III. 4).

"Related to *E. hirta* and *E. parviflora* but yet sufficiently different from these. S."

Jatropha curcas *L.*—FWTA. ed. 2, 1: 397 (1958); Junghans 1961: 337.

J. curcas L. var. *β glabrata* Schum. et Thonn. in Schum. 1827: 412 and 1829: 186.

Exsicc.: "Guinea", *Thonning* 110 (**C**—2 sheets, types, microf. 61: II. 7; III 1–3).

"Here and there."
"Specimens from East Indies differ from our plant in having rather shorter flower-stalks and a rust coloured down on calyx and leaflet-stalks. West Indian specimens approach nearer to our plant in smoothness of their parts. S."

Note: Although of doubtful taxonomic value, this variety seems to have been overlooked.—F.N.H.

Mallotus oppositifolius (*Geisel.*) *Müll. Arg.*—FTA. 6, 1: 928 (1912); FWTA. ed. 2, 1: 402 (1958).

Croton oppositifolius Geisel. 1807: 23 ("Habitat in Guinea ex herb. *Iserti*"); Junghans 1961: 336.

Exsicc.: Whydah, 1785, *Isert* s.n. (**C**—2 sheets, type, microf. 28: II. 7; III. 1, 2).

Acalypha? dentata Schumacher 1827: 410 and 1829: 184 (no MS.).

Exsicc.: "Guinea", *Thonning* s.n. (**C**—2 sheets, types, microf. 1: I. 3–7); "e Guinea" (**C**—1 sheet, microf. 1: II. 1, 2; **P–JU**—1 sheet, No. 16585, microf. 1200: I. 5).

"Here and there."
"The flowers smell like *Convallaria majalis*. Th."
"I leave it to others to decide, whether it belongs to this genus [i.e *Acalypha*] or should constitute a separate one. S."

Note (1): Junghans cited 3 sheets "e Guinea" but one of them bears Thonning's initials and another (microf. 1: II. 3, 4) is *Amyris sylvatica* Jacq. (Rutaceae) of non-African origin—F.N.H.

Note (2): There is a small packet on *Welwitsch* 342 of this species at Geneva labelled simply "Guinea", which might be a Thonning specimen.—F.N.H.

Manihot esculenta Crantz—FWTA. ed. 2, 1: 413 (1958).

Janipha manihot (L.) Kunth—Schum. 1827: 414 and 1829: 188 (no MS); Junghans 1961: 337.

Exsicc.: "Guinea", *Thonning* s.n. (**C**—2 sheets, microf. 65: I. 6, 7; II: 3, 4); "e Guinea?" (**C**—1 sheet, microf. 65: II. 1, 2).

"Cultivated".

"The Guinea plant differs from the Indian in its smaller leaves, which above are darker, below nearly light blue; and in that the root is not poisonous. S."

Microdesmis puberula *Hook. f. ex Planch.*—FWTA. ed. 2, 1: 392 (1958).

Euphorbiaceae indet. in Junghans 1961: 338.

Exsicc.: "Guinea", *Thonning* s.n. (**C**—1 sheet, microf. 84: III. 3, 4).

NOTE: This genus is now usually placed in Pandaceae—F.N.H.

Phyllanthus amarus *Schum. et Thonn.* in Schum. 1827: 421 and 1829: 195; FWTA. ed. 2, 1: 387 (1958); Junghans 1961: 337.

Exsicc.: "Guinea", *Thonning* 4 (**C**—4 sheets, types, microf. 80: III. 2–6; 83: III. 3, 4); "e Guinea" (**C**—1 sheet, microf. 81: I. 1, 2; **FI**—1 sheet).

VERNACULAR NAME: 'Aumaadoati'.

"Frequent".

"Very rarely 3 flowers are seen together of which 2 are male flowers. When the plant is dying not only the leaves but very often the leaf-stalks, the fruit and the entire upper part of the plant are downy and the stipules are rust coloured. It grows in most places and in every kind of soil; after heavy rain it frequently turns up in cultivated places, but as in such places there is no grass which can shade it from the burning sun it rarely attains its full growth as by the R. Volta and some other places, where it, shaded by the tall grass, becomes a little shrub. As soon as ever the plant has leaves it also has flowers and it requires a very short period for growing to 1 ft. in height and bearing a number of seeds. I tried to sow the seed in good soil, where it itself frequently appears elsewhere, and did not neglect to water it but not one plant came up even after 4 weeks. Its leaves have a penetrating pure bitter taste mainly after being dried; the herbaceous plant is superior, in this respect far surpassing *Menyanthes trifoliata* in bitterness and pure taste; with brandy it yields a very bitter essence: the natives boil it with Pytto and use it in this way for expelling fever and stomach pains. For dropsy both in the abdomen and beneath the skin I have seen it used with much efficacy; the plant is crushed along with a few grains-of-Paradise [*Aframomum*] and rubbed in over the whole body; a decoction of the plant is used internally. Th."

"It differs from *Phyllanthus niruri* in its streaked stem and in the flowers placed 2 and 2 together; perhaps however it is only a subspecies. S."

Phyllanthus capillaris *Schum. et Thonn.* in Schum. 1827: 417 and 1829: 191; FWTA. ed 2, 1: 387 (1958); Junghans 1961: 337. *Menarda capillaris* (Schum. et Thonn.) Baill. 1860: 85.

Exsicc.: Aquapim, 1786, *Isert* s.n. (**C**—1 sheet, type, microf. 81: II. 2);

Aquapim, *Thonning* 159 (**C**—5 sheets, types, microf. 81: II. 3–7; III. 1–5); "e Guinea" (**FI**—1 sheet; **G–DC**—1 sheet, 15(2). 338. 179, microf. 2478: I. 8; **P–JU**—1 sheet, No. 16303, microf. 1180: III. 4).

"Grows only in damp and rich soil, preferably in thickets in and around Aquapim."

Phyllanthus maderaspatensis *L.*—FTA. 6, 1: 722 (1912); FWTA. ed. 2, 1: 388 (1958).

P. thonningii Schumacher 1827: 418 and 1829: 192; Junghans 1961: 338.

EXSICC.: Prampram & Christiansborg, *Thonning* 27 (**C**—4 sheets, types, microf. 84: II. 3–7; III. 1, 2); "e Guinea" (**G–DC**—1 sheet, 15(2): 362. 243, microf. 2482: I. 4; **P–JU**—1 sheet, No. 16238, microf. 1176: III. 3).

"Grows only at Prampram and Christiansborg in a few places."
"As regards the branches it comes near to *P. anceps* but in the latter the leaves are smaller and are placed closer together; the flowers singly in the leaf-axils, flower-stalks longer hair-like; capsules much smaller and finally the branches two-edged. S."

Phyllanthus pentandrus *Schum. et Thonn.* in Schum. 1827: 419 and 1829: 193; FWTA. ed. 2, 1: 387 (1958); Junghans 1961: 337. *Menarda linifolia* Baill. 1860: 84.

EXSICC.: Christiansborg, *Thonning* 34 (**C**—6 sheets, types, microf. 82: III. 5, 6; 83: I. 1–7; II. 1, 4, 5); "e Guinea" (**C**—2 sheets, microf. 82: III. 4; 83: II. 2, 3; **FI**—1 sheet; **G–DC**—1 sheet, 15(2): 337. 177, microf. 2478: I. 5; **P–JU**—1 sheet, No. 16239, microf. 1176: III. 4).

"Rare; near the forts Christiansborg and Prindsensteen and near Dudua."

NOTE: Vahl gave his specimens the MS name *Phyllanthus linifolius*, on which Baillon based his *Menarda linifolia.*—F.N.H.

Phyllanthus reticulatus *Poir.* var. **glaber** *Müll. Arg.*—FTA. 6, 1: 701 (1912); FWTA. ed. 2, 1: 387 (1958).

P. polyspermus Schum. et Thonn. in Schum. 1827: 416 and 1829: 190; Junghans 1961: 337.

EXSICC.: Quitta, *Thonning* 51 (**C**—2 sheets, types, 83: II. 6, 7; III. 1, 2).

"At Quitta, rare."
"Sufficiently different from *P. multiflorus* to which it approaches. S."

Phyllanthus sublanatus *Schum. et Thonn.* in Schum. 1827: 420 and 1829: 194; FWTA. ed. 2, 1: 387 (1958); Junghans 1961: 337.

EXSICC.: "Guinea", *Thonning* 37 (**C**—6 sheets, types, microf. 83: III. 5–7; 84: I. 1–5, II. 1, 2); "e Guinea" (**C**—1 sheet, microf. 84: I. 6, 7; **P–JU**—1 sheet, No. 16305, microf. 1180: III. 6).

"Here and there, rarer than the following one [i.e. *P. amarus*]."

"The bitter taste which the following species [i.e. *P. amarus*] has, is not found in this one; the natives distinguish it therefrom by the green leaf-stalk, which however is not a sure character. The older the plant is the more downy are the leaves; the young plant is often quite naked. Th."

"From the allied species: *P. debilis, niruri, urinaria* etc. it differs in its small leaves and the rest of the cited characters. S."

Securinega virosa (*Roxb. ex Willd.*) *Baill.*—FWTA. ed. 2, 1: 389 (1958). *Fluggea microcarpa* Blume—FTA. 6, 1: 736 (1912).

Phyllanthus angulatus Schum. et Thonn. in Schum. 1827: 415 and 1829: 189; Junghans 1961: 337.

Exsicc.: "Guinea," *Thonning* 33 (**C**—3 sheets, types, microf. 81: I. 3–7; II. 1); "e Guinea" (**P–JU**—1 sheet, No. 16333, microf. 1182: III. 2).

Vernacular name: 'Lomo-tjo'.

"Common on slopes in neighbourhood of the sea."

"In habit much resembles *P. virosus* but in the latter the leaves are elliptical-ovate, the older ones pointed, the younger most frequently excised without mucro, underneath net-veined; the twigs are open and greyish. Thonning does not state whether the seeds are poisonous. S."

P. dioicus Schumacher 1827: 416 and 1829: 190 (no MS.)

Exsicc.: "Guinea", 1785, *Isert* s.n. (**C**—1 sheet, microf. 82: II. 1, 2); "Guinea", *Thonning* s.n. (**C**—8 sheets, types, microf. 82: I. 1–7; II. 3–7; III. 1–3); "e Guinea" (**C**—1 sheet, microf. 81: III. 6).

"Grows with the foregoing [i.e. *P. angulatus*]."

"Certainly different from the preceding species in the shape of the leaves, grey colour of the branches and the whole habit. It is also different from *P. virosus*. S."

Note: A sheet in Willdenow's Herbarium No. 17964 labelled *Phyllanthus virosus* [*P. viscosus* by error in the Index] has the locality "Guinea" written on it, although "India orientale" appears on the label. There is no indication of the collector.—F.N.H.

Tragia monadelpha *Schum. et Thonn.* in Schum. 1827: 404 and 1829: 178; FWTA. ed. 2, 1: 412 (1958); Junghans 1961: 338.

Exsicc.: Aquapim, *Thonning* 222 (no type material traced).

"In and near Aquapim."

"Should perhaps constitute a separate genus owing to the four-leaved calyx, the connate stamen-filaments with about 4 anthers in the male-flowers; the five-lobed calyx, absence of a style and the penicillate stigma in the female flower. S."

Note: This remains an imperfectly known species.—F.N.H.

FICOIDACEAE see AIZOACEAE

FLACOURTIACEAE

Flacourtia flavescens *Willd.* 1806a: 830 ("Habitat in Guinea. *Isert*");
FTA. 1: 121 (1868); FWTA. ed. 2, 1: 189 (1954); Junghans 1961: 338.

Exsicc.: "Guinea", *Isert* s.n. (**B**—1 sheet, No. 18491, type, microf. 1344:
II. 7).

F. edulis Schum. et Thonn. in Schum. 1827: 450 and 1829: 224.

Exsicc.: "Guinea", *Thonning* 55 (**C**—1 sheet, type, microf. 42: III. 7);
"e Guinea" (**P–JU**—1 sheet, No. 12594, type, microf. 931: II. 2).

Vernacular name: 'Amagomi'.

"Here and there."

"Perhaps it is the same plant as *F. flavescens* Willd. but the very imperfect
description does not permit of its being determined with certainty. The
leaves are described as being oblong, obtuse, narrower towards the base;
whereas in our plant they are ovate, acuminate, broad at the back. S."

"Near the coast they are seldom more than 2 ft. high; below the mountains
on the other hand and on the mountains themselves they are much over a
man's height, and the trunk attains the thickness of an arm. The wood is
reddish, hard and fine and very twisted. The berries are among the best
wild fruits, the pulp is sweetish, mealy, and can be enjoyed in as large a
quantity as one likes without harm. The berry which grows near the coast is
far more pleasing than that which comes from the fertile mountains; the
same is the case with most fruits, such as pineapples, oranges, guavas etc.,
they are much smaller but much sweeter and richer in taste. The young
leaves are put in pytto which is set in the sun to become sour, and is drunk
thus in cases of gonorrhea for expelling urine. In wasting cough a portion of
the leaves is taken and some malage [? melagueta, *Aframomum*] in the mouth
and is chewed; the juice is swallowed with cold water. Th."

Oncoba spinosa *Forsk.*—FTA. 1: 115 (1868); FWTA. ed. 2, 1: 188 (1954).

Lundia monacantha Schum. et Thonn. in Schum. 1827: 231 and 1829: 5;
Junghans 1961: 338.

Exsicc.: *Thonning* 296 (no type material traced).

Vernacular name: 'Azara-tjo'.

"Rare"

"Perhaps it is the same plant which Forskaal described under the name of
Oncoba (centur. 4 p. 103 no. 21); in habit it agrees well with the plant here
described but the thorns are shorter (altho' Forskaal ascribes to them a
length of 2 inches), erect not horizontal, and the berry unilocular. S."

"The natives clean out the fruit and use it then as a snuff box; they merely
insert a stopper in it. The plant flowers in May. Th."

"I have called this plant after the late titular state councillor Tönder
Lund, deputy in the General Custom house, known through his treatises

on natural history which are printed in the publications of the Copenhagen
Natural History Society. S."

GOODENIACEAE

Scaevola plumieri (*L.*) *Vahl*—FWTA. ed. 2, 2: 315 (1963).

Scaevola lobelia Murr. 1774: 178; Schum. 1827: 106 and 1828: 126; Junghans
1961: 339.

EXSICC.: *Thonning* 22 (no material traced).

VERNACULAR NAME: 'Gubaa'.

"Everywhere along shore-margins or salt lagoons, where it serves to bind
the loose shore-sand."

HIPPOCRATEACEAE see CELASTRACEAE

HYDROPHYLLACEAE

Hydrolea glabra *Schum. et Thonn.* in Schum. 1827: 161 and 1828: 181;
FWTA. ed. 2, 2: 316 (1963); Junghans 1961: 339.

EXSICC.: Aquapim, *Thonning* 198 (**C**—1 sheet, type, microf. 54: III. 3, 4);
"e Guinea" (**C**—1 sheet, microf. 54: III. 2; **P–JU**—1 sheet, No. 6913,
microf. 506: II. 6).

"Common in boggy and shady places in Aquapim."
"*H. zeylanica* differs in having linear-lanceolate almost sessile leaves and
long racemes which occur in the leaf-axils between the leaves. We have
received from the E. Indies a plant under this name, which is probably
different therefrom, as the flower-stalks and calyx are beset with gland-
bearing hairs; whereas these parts are smooth in *H. zeylanica* (*Nama zeylanica*
Linn.) S."

LABIATAE

Basilicum polystachyon (*L.*) *Moench*—FWTA. ed. 2, 2: 454 (1963).

Ocymum dimidiatum Schum. et Thonn. in Schum. 1827: 267 and 1829: 41;
Junghans 1961: 339.

EXSICC.: "Guinea", *Thonning* 309 (**C**—2 sheets, types, microf. 71: II. 2–5).

"Here and there."
"It differs from *O. polystachium* in its much longer and thinner racemes, in
more distant whorls, which only occupy half the circumference, and in less
stiff-haired calyx and rachis. S."

Endostemon tereticaulis (*Poir.*) *M. Ashby*—FWTA. ed. 2, 2: 453 (1963).
Ocymum tereticaule Poir.—FTA. 5: 347 (1900); Junghans 1961: 340.

O. thonningii Schumacher 1827: 265 (non 269) and 1829: 39 (non 43).

EXSICC.: "Guinea", *Thonning* 78 (**C**—2 sheets, type, microf. 72: II. 4–7; **FI**—1 sheet, type; **S**—1 sheet, type); "e Guinea" (**C**—1 sheet, microf. 72: III. 1, 2; **P–JU**—1 sheet, No. 5694, microf. 406: III. 6).

NOTE: See comment on the name *O. thonningii* at the top of p. 66.

"Here and there. Flowers in June."

Hoslundia opposita *Vahl et Thonn.* in Vahl 1804: 212; Schum. 1827: 15 and 1828: 35 (as *Haaslundia*); FWTA. ed. 2, 2: 516 (1963); Junghans 1961: 339.

EXSICC.: Aquapim, *Thonning* 196 (**C**—3 sheets, type, microf. 48: III. 3–6; 49: I. 1, 2; **S**—1 sheet, type); "e Guinea", **P–JU**—1 sheet, No. 5272, microf. 370: II. 7).

"Amongst bushes in and near Aquapim."
"When naming after persons one ought not to change the name of the latter and Thonning's friend and travelling companion bore the name of Ole Haaslund Smith, I have named the genus "Haaslundia" instead of Vahl's name "Hoslundia.""

Hyptis lanceolata *Poir.*—FWTA. ed. 2, 2: 466 (1963). *H. brevipes* sensu FTA. 5: 447 (1900).

H. lanceifolia Thonning in Schum. 1827: 261 and 1829: 35; Junghans 1961: 339.

EXSICC.: Ada, 1784, *Isert* s.n. (**C**—1 sheet, microf. 55: I. 4); Aquapim, *Thonning* 224 (**C**—3 sheets, types, microf. 55: I. 6, 7; II. 1–4); "e Guinea?" (**C**—1 sheet, microf. 55: I. 5; **P–JU**—1 sheet, No. 5412, microf. 382: II. 1; **FI**—1 sheet).

"Not common near Aquapim."

Hyptis pectinata (*L.*) *Poit.*—FWTA. ed. 2, 2: 466 (1963).

Bystropogon coarctatus Schum. et Thonn. in Schum. 1827: 260 and 1829: 34; Junghans 1961: 339.

EXSICC.: Ningo, *Thonning* 320 (**C**—3 sheets, types, microf. 13: I. 2–6).

"At Ningo."
"Comes near to *B. pectinatus* but in the latter the racemes are longer, looser, the whorls pectinate, more distant, teeth of the calyx longer, calyx and stem slightly hairy, the teeth of the leaves larger and more acute. S."

Labiatae indet. in Junghans 1961: 340.

EXSICC.: "Guinea", *Mortensen* s.n. (**C**—1 sheet, microf. 55: II. 5, 6).

Leonotis nepetifolia (*L.*) *Ait. f.* var. **africana** (*P. Beauv.*) *J. K. Morton*—FWTA. ed. 2, 2: 470 (1963).

Phlomis pallida Schum. et Thonn. in Schum. 1827: 262 and 1829: 36; Junghans 1961: 340.

Exsicc.: Quitta and Ursue, *Thonning* 1 (**C**—1 sheet, type, microf. 80: I. 3, 4).

"Near Quitta and Ursue. Flowers almost the whole year through."

"Comes nearest to *P. nepetaefolia* but in the latter the furrows of the stem are more streaked, the leaf-stalks longer than the leaves, the calyx 6–8 toothed, the uppermost and lowest tooth larger; the corolla is double as long, orange-coloured not whitish-yellow. S."

"Its odour comes near to that of *Marrubium vulgare*. A decoction of the dried plant is used as a drink in cases of persistent coughing. Th."

Leucas martinicensis *(Jacq.) Ait. f.*—FWTA. ed. 2, 2: 470 (1963).

Phlomis mollis Schum. et Thonn. in Schum. 1827: 263 and 1829: 37; Junghans 1961: 340.

Exsicc.: Whydah, 1783, *Isert* s.n. (**C**—1 sheet, microf. 79: III. 2); "Guinea", *Thonning* 310 (**C**—4 sheets, type, microf. 79: III. 3–7; 80: I. 1, 2; **S**—1 sheet, type); "e Guinea" (**FI**—1 sheet; **P–JU**—1 sheet, No 5614A, type, microf. 400: I. 4).

"Here and there."

"*P. caribea* v. *martinicensis* much resembles it, but differs in having lanceolate leaves with smaller serrations and without hairs on the upper side, and a more deeply-furrowed stem, 8-toothed calyx, and a corolla which is longer than the calyx. S."

NOTE: The FI specimen has a label by Vahl '*Phlomis breviflora*' which is a MS name—F.N.H.

[Ocimum barbatum *Vahl* in Schum. 1827: 267 and 1829: 41.

Exsicc.: India, *König* s.n. (**C**—1 sheet, microf. 72: I. 5, 6).

NOTE: This Indian species is only included here as it is discussed by Schumacher and appears in the microfiche—F.N.H.]

Ocimum canum *Sims*—FWTA. ed. 2, 2: 452 (1963).

Exsicc.: "e Guinea" (**P–JU**—1 sheet, No. 5695, microf. 406: III. 7).

O. hispidulum Schum. et Thonn. in Schum. 1827: 266 and 1829: 40; Junghans 1961: 340.

Exsicc.: "Guinea", *Thonning* 295 (**C**—4 sheets, types, microf. 71: III. 1–7; **FI**—1 sheet).

VERNACULAR NAME: 'Koae'.

"Here and there."

"Comes near *O. pilosum*, *O. ciliatum* Horn. Hort. and *O. barbatum* Herb. Vahl but is yet different from these species. It is doubtful whether these plants are really different species. S."

"The whole plant has an extremely strong and fairly pleasant odour; it is used by the natives in various illnesses, mainly such as are attributed to witchcraft or the deceased. Th."

(a) A Thonning sheet of *Abrus precatorius* L. at Copenhagen.

(b) A Thonning sheet of *Indigofera pilosa* Poir. at Stockholm.

PLATE 4

(b) A Mortensen sheet of *Microlepia speluncae* (L.) Copel. at Copenhagen.

(a) A Thonning sheet of *Indigofera pilosa* Poir. (note small label) at Florence.

PLATE 5

BESKRIVELSE

\ !'

GUINEISKE PLANTER

SOM ERE FUNDNE

AF DANSKE BOTANIKERE, ISÆR AF

ETATSRAAD *THONNING*

VED

F. C. SCHUMACHER,

PROFESSOR VED UNIVERSITETET I KIÖBENHAVN,

RIDDER AF DANNEBROGEN &c.

Særskilt aftrykt af det kongelige danske Videnskabers Selskabs Skrifter.

KJÖBENHAVN, 1827.

TRYKT I HARTV. FRID. POPPS BOGTRYKKERIE.

Title page of F. C. Schumacher's *Beskrivelse* showing the controversial date 1827 of the pre-print with consecutive pagination.

PLATE 6

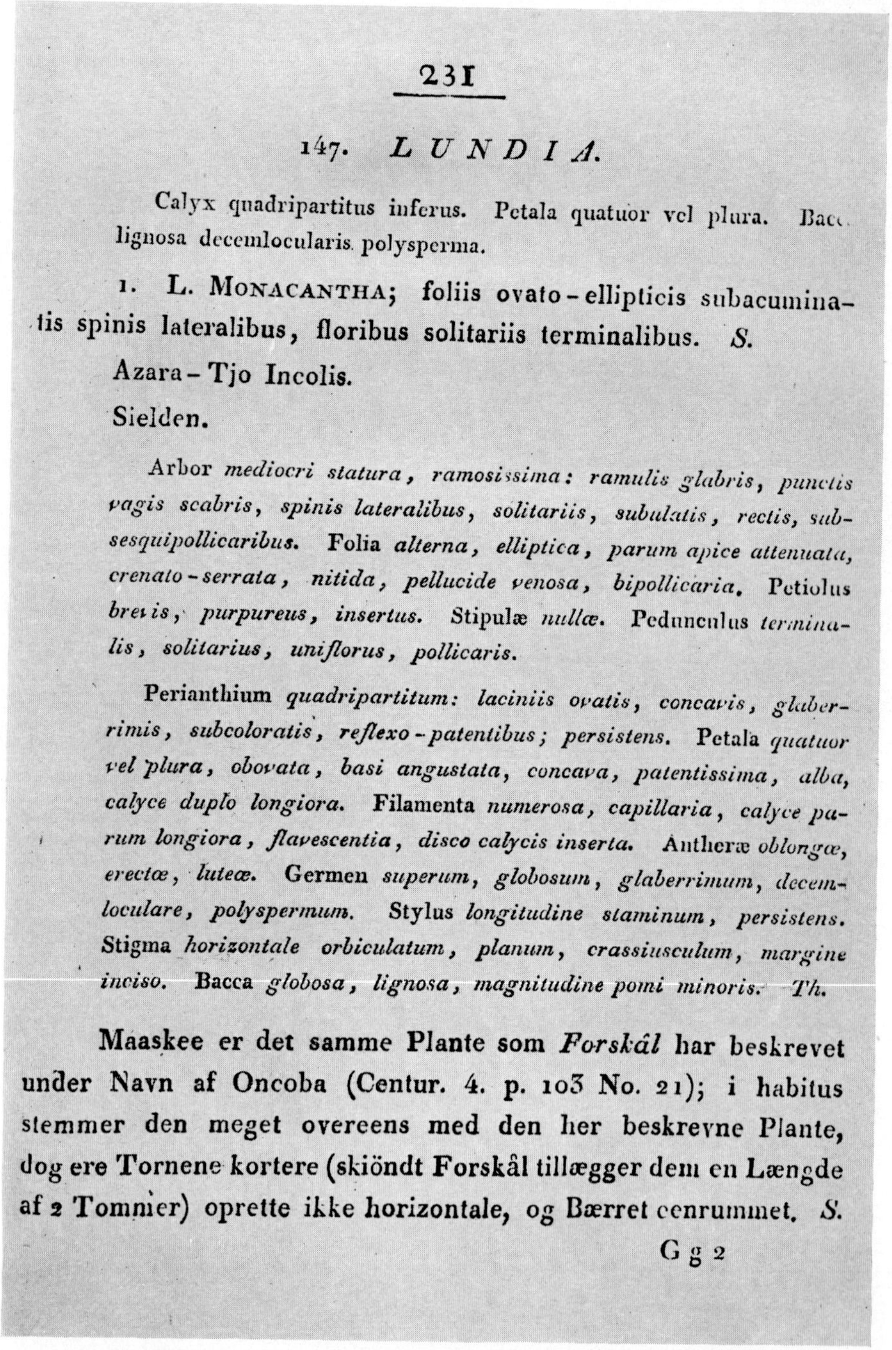

147. *L U N D I A*.

Calyx quadripartitus inferus. Petala quatuor vel plura. Bacc. lignosa decemlocularis. polysperma.

1. L. MONACANTHA; foliis ovato-ellipticis subacumina-tis spinis lateralibus, floribus solitariis terminalibus. *S.*

Azara - Tjo Incolis.

Sielden.

Arbor *mediocri statura, ramosissima: ramulis glabris, punctis vagis scabris, spinis lateralibus, solitariis, subulatis, rectis, subsesquipollicaribus.* Folia *alterna, elliptica, parum apice attenuata, crenato-serrata, nitida, pellucide venosa, bipollicaria.* Petiolus *brevis, purpureus, insertus.* Stipulæ *nullæ.* Pedunculus *terminalis, solitarius, uniflorus, pollicaris.*

Perianthium *quadripartitum: laciniis ovatis, concavis, glaberrimis, subcoloratis, reflexo-patentibus; persistens.* Petala *quatuor vel plura, obovata, basi angustata, concava, patentissima, alba, calyce duplo longiora.* Filamenta *numerosa, capillaria, calyce parum longiora, flavescentia, disco calycis inserta.* Antheræ *oblongæ, erectæ, luteæ.* Germen *superum, globosum, glaberrimum, decemloculare, polyspermum.* Stylus *longitudine staminum, persistens.* Stigma *horizontale orbiculatum, planum, crassiusculum, margine inciso.* Bacca *globosa, lignosa, magnitudine pomi minoris. Th.*

Maaskee er det samme Plante som *Forskål* har beskrevet under Navn af Oncoba (Centur. 4. p. 103 No. 21); i habitus stemmer den meget overeens med den her beskrevne Plante, dog ere Tornene kortere (skiöndt Forskål tillægger dem en Længde af 2 Tommer) oprette ikke horizontale, og Bærret eenrummet. *S.*

Gg 2

A page of F. C. Schumacher's *Beskrivelse* to show the lay-out. Note initials S. and Th.

PLATE 7

O. lanceolatum Schum. et Thonn. in Schum. 1827: 268 and 1829: 42.

EXSICC.: "Guinea", *Thonning* 31 (**C**—2 sheets, types, microf. 72: I. 1–4; **S**—1 sheet, type).

VERNACULAR NAME: 'Blafaa-koae'.

"This plant is imported, but it multiplies itself where it has once been sown."

"Comes close to *O. basilicon* but in the latter the bracts are less hairy on the margin, the leaf-stalks smooth, the internodes downy-haired and not bristly on the edges, the leaves more ovate oblong. S."

"The natives have the superstition that the spirit of their deceased relatives can appear to them and bring upon them all kinds of illness; when a native attributes his illness to this cause, this and other strongly-scented plants are boiled in water, with which he washes himself and one besprinkles the place where he is staying with the powerful odour in order to expel the spirit. Th."

Labiatae indet. in Junghans 1961: 340.

EXSICC.: Ada, 1784, *Isert* s.n. (**C**—1 sheet, microf. 71: I. 3).

Ocimum gratissimum *L.*—FWTA. ed. 2, 2: 452 (1963).

O. guineense Schum. et Thonn. in Schum. 1827: 264 and 1829: 38; Junghans 1961: 339.

EXSICC.: Ada, *Thonning* 23 (**C**—1 sheet, type, microf. 71: II. 6, 7).

VERNACULAR NAME: 'Sylu' (Accra dialect)

"Grows frequently in neighbourhood of Ada. In some places near the coast some few plants are grown for the sake of medicinal use, but these are in general stunted."

"For most of their illnesses the natives take various baths; most of these are connected with fetish and work only along with a superstitious faith, but some plants, e.g. the present, act with actual aromatic ingredients. The most important use of this plant is in a malignant bilious fever connected with jaundice, which is very prevalent near R. Volta after inundation of the river. One uses generally, the tepid decoction both for drinking and bathing 4 times a day. It is used in the same way in common jaundice. In sudden frenzy or swooning without preceding illness one drips the expressed sap into the nose, eyes and mouth for expelling the sirsa (spirit or ghost) which is thought to have attacked the sick person. I have successfully treated old bone-injury and strongly running eruptions or the so-called salt flux by the external use of a decoction of the unripe fruit of *Hibiscus esculentus* together with this plant [i.e. *Ocymum gratissimum*] and laxative agents. Th."

Orthosiphon suffrutescens (*Schumacher*) *J. K. Morton*—FWTA. ed. 2, 2: 454 (1963).

Ocymum suffrutescens Schumacher 1829: 330, 335 (see note); Junghans 1961: 340.

EXSICC.: "Guinea", *Thonning* 288 (**C**—2 sheets, microf. 72: I. 7, II. 1–3).

c

O. thon(n)ingii Schumacher 1827: 269 (non 265) and 1829: 43 (non 39), see note.

VERNACULAR NAME: 'Sissa-koae'.

"Here and there."

NOTE: In 1829 *O. suffrutescens* was substituted for *O. thon(n)ingii*, which had been used already for another species, *Endostemon tereticaulis.*—F.N.H.

Platostoma africanum *P. Beauv.*—FWTA. ed. 2, 2: 453 (1963).

Ocymum sylvaticum Thonning in Schum. 1827: 270 and 1829: 44; Junghans 1961: 340.

EXSICC.: Aquapim, *Thonning* 223 (no type material traced).

"Common in woods in Aquapim."

Solenostemon monostachyus (*P. Beauv.*) *Briq.* subsp. **monostachyus**— FWTA. ed. 2, 2: 464 (1963).

S. ocymoides Schum. et Thonn. in Schum. 1827: 271 and 1829: 45; Junghans 1961: 340.

EXSICC.: Whydah, 1782, *Isert* s.n. (**C**—1 sheet, type, microf. 102: III. 6); "Guinea", *Thonning* 92 (**C**—6 sheets, type, microf. 102: III. 7; 103: I. 1–7, II. 1–5); "e Guinea" (**P–JU**—1 sheet, No. 5704, microf. 407: III. 1).

VERNACULAR NAME: 'Keriro'.

"Here and there. Flowers in June."
"Has the appearance of *Ocymum* and *Scutellaria*. Name from two Greek words meaning tube and stamen."

LAURACEAE

Cassytha filiformis *L.*—FTA. 6, 1: 188 (1909); FWTA. ed. 2, 1: 58 (1954).

Cassyta guineensis Schum. et Thonn. in Schum. 1827: 199 and 1828: 219; Junghans 1961: 340.

EXSICC.: "Guinea", 1785, *Isert* s.n. (**C**—1 sheet, microf. 16: II. 6); "Guinea", *Thonning* 8 (**C**—3 sheets, types, microf. 16: I. 7; II. 1–5).

"It resembles *Volutella* Forsk. and *C. filiformis* from India; but in these the stems and flower-stalks are smooth, the racemes longer and the flowers are more distant from each other. In a specimen from the Cape the stems are smooth and the flower-stalks downy-hairy, 2–3–4 inches long, curved, the flowers more distant. It appears therefore that our plant should be distinguished as a separate species. S."

LENTIBULARIACEAE

Utricularia inflexa *Forsk.* var. **inflexa**—Vahl 1804: 196, as to plant from Guinea; FWTA. ed. 2, 2: 380 (1963).

U. thonningii Schumacher 1827: 12 and 1828: 32; Junghans 1961; 341.

Exsicc.: "Guinea", *Thonning* 331 (**C**—1 sheet, type, microf. 108: III. 4); "e Guinea" (**FI**—1 sheet; **P–JU**—1 sheet, No. 4828, microf. 331: II. 7).

"In stagnant water but rare."

"Vahl assumed in his Enumeration that this plant was the same as Forskaal described under the name of *U. inflexa*; on the contrary it appears to me different therefrom. In *U. inflexa* the utriculi on the scape are ovate, and have a lengthened tip, which is provided with a number of filaments, the nectary is bent upwards, truncate at the end, and felted as Forskal says in Flora Aegyptico-arabica p. 10. Besides the flowers in *U. inflexa* are larger, the leaf-baring roots much shorter, utriculi much fewer, even quite wanting sometimes; and this is doubtless why Vahl states: variat cum et absque bullis in foliis radicalibus. S."

Note: Junghans apparently did not find Thonning's specimen cited above as he noted that there was no material of it at Copenhagen.—F.N.H.

LOGANIACEAE

Usteria guineensis *Willd.* 1792: 55, pl. 2; 1797a: 18 ("Habitat in Guinea"); Vahl 1804: 5 ("Habitat in Guinea. *Isert*"); Schum. 1827: 7 and 1828: 27 (no MS); FWTA. ed. 2, 2: 46, fig. 210 (1963); Junghans 1961: 341.

Exsicc.: Whydah, 1785, *Isert* s.n. (**B**—1 sheet, No. 47, type, microf. 3: II. 8; **C**—1 sheet, type, microf. 108: III. 3; **M**—1 sheet, type).

"In Guinea according to Isert's herbarium. Vahl."

LORANTHACEAE

Phragmanthera incana (*Schumacher*) *Balle*—FWTA. ed. 2, 1: 664 (1958). *L. thonningii* DC. 1820: 303, non Schumacher.

Loranthus incanus Schum. et Thonn. in Schum. 1827: 180 and 1828: 200; FTA. 6, 1: 292 (1910); Junghans 1961: 341.

Exsicc.: Ada, 1784, *Isert* s.n. (**C**—1 sheet, microf. 65: I. 3); "Guinea", *Thonning* 321 (**C**—1 sheet, type, microf. 65: I. 4; **G–DC**—1 sheet, 4: 303. 125, type, microf. 669: II. 1).

"At the river Volta, not common."

Tapinanthus bangwensis (*Engl. et K. Krause*) *Danser*—FWTA. ed. 2, 1: 662 (1958).

Loranthus thonningii Schumacher 1827: 179 and 1828: 199; FTA. 6, 1: 363 (1910); Junghans 1961: 341.

Exsicc.: *Thonning* 126 (no type material traced).

Vernacular name: 'Eduásudoá'.

"Common on various trees."

"At times it is used in various illnesses to consecrate water, with which the sick person may be washed. It is besides a harmful parasite especially for fruit trees. Th."

"Seems to come near *L. sessilifolius* P. Beauv. (Flore d'Ovare et de Benin. Tab. LXIII)."

NOTE: Junghans says it may be found elsewhere as *L. umbellatus* Vahl, but it is not in the Willdenow Herbarium under that name. The synonymy suggested by Sprague in FTA. from the description needs to be confirmed.—F.N.H.

MALPIGHIACEAE

Acridocarpus alternifolius *(Schum. et Thonn.) Niedenzu*—FWTA. ed. 2, 1 : 352 (1958).

Malpighia alternifolia Schum. et Thonn. in Schum. 1827: 222 and 1828: 242; Junghans 1961: 342.

EXSICC.: "Guinea", *Thonning* 173 (**C**—1 sheet, type, microf. 65: I. 5); "e Guinea" (**P–JU**—1 sheet, No. 11661, microf. 866: I. 3).

"Rare. In a copse at Kratté."

Triaspis odorata *(Willd.) A. Juss.*—FTA. 1: 280 (1910); FWTA. ed. 2, 1: 354 (1958).

Hiraea odorata Willd. 1799: 743 ("Habitat in Guinea. *Isert*"); Schum. 1827: 223 and 1828: 243 (no MS); Junghans 1961: 341.

EXSICC.: Ada, 1787, *Isert* s.n. (**B**—1 sheet, No. 8861, type, microf. 614: II. 7; **C**—1 sheet, type, microf. 54: II. 3); "Guinea", *Thonning* s.n. (**C**—1 sheet, microf. 54: II. 2); "e Guinea"? (**C**—2 sheets, microf. 54: II. 4, 5; **P–JU**—1 sheet, No. 11675, microf. 867: I. 2).

"At Ada."

MALVACEAE

Abelmoschus esculentus *(L.) Moench.*—Hauman in Fl. Congo 10: 143 (1963).

Hibiscus esculentus L.—Schum. 1827: 316 and 1829: 90 (no MS.); FWTA. ed. 2, 1 : 348 (1958); Junghans 1961: 342.

EXSICC.: "Guinea", *Thonning* s.n. (**C**—1 sheet, microf. 53: I. 5, 6).

"Cultivated here and there."

NOTE: Schumacher's var. β with deeply lobed leaves remained un-named.—F.N.H.

Abelmoschus moschatus *Medik.*—Hauman in Fl. Congo 10: 144 (1963).

Hibiscus abelmoschus L.—Schum. 1827: 315 and 1829: 89; FWTA. ed. 2, 1: 347 (1958); Junghans 1961: 342.

EXSICC.: "Guinea", *Thonning* 112 (**C**—2 sheets, microf. 52: III. 4, 5, only 1 sheet photographed).

VERNACULAR NAME: 'Asianté-Kitteva'.

"Cultivated here and there but in small quantity."

Abutilon guineense (*Schum. et Thonn.*) *Bak. f. et Exell*—FWTA. ed. 2, 1: 337 (1958).

Sida guineensis Schum. et Thonn. in Schum. 1827: 307 and 1829: 81; Junghans 1961: 343.

EXSICC.: "Guinea", *Thonning* 38 (**C**—5 sheets, types, microf. 99: I. 5-7, II. 1, 2, only 3 sheets photographed; **S**—1 sheet, type).

"Here and there."
"Seems to me different from *Sida asiatica* which it otherwise approaches; leaves in the latter are pointed, dissimilar and with sharp serrations; fruit nodding shorter than the calyx. S."

Gossypium barbadense L.—FTA. 1: 210 (1868); FWTA. ed. 2, 1: 349 (1958).

G. punctatum Schum. et Thonn. in Schum. 1827: 309 and 1829: 83; Junghans 1961: 342.

EXSICC.: *Thonning* 327 (no type specimen traced).

"Cultivated".
"Appears nearly related to *G. religiosum*. S."

Gossypium herbaceum *L.* var. **acerifolium** (*Guill. et Perr.*) *A. Chev.*— FWTA. ed. 2, 1: 349 (1958).

G. prostratum Schum. et Thonn. in Schum. 1827: 311 and 1829: 85; Junghans 1961: 342.

EXSICC.: *Thonning* 328 (no type specimen traced at **C**; **S**—1 sheet, type).

"Cultivated".

NOTE: Thonning noted a small variety without giving it a name.—F.N.H.

Hibiscus cannabinus *L.*—FTA. 1: 204 (1868); FWTA. ed. 2, 1: 347 (1958).

H. congener Schum. et Thonn. in Schum. 1827: 319 and 1829: 93; Junghans 1961: 342.

EXSICC.: L. Augna, *Thonning* 130 (**C**—3 sheets, types, microf. 52: III. 6, 7; 53: I. 1-4; **S**—1 sheet, type).

VERNACULAR NAME: 'Sissa-imune'.

'Here and there in marshy places, not very common e.g. on the smaller islands in the Augna Lake."

"In appearance it much resembles *H. cannabinus* but it seems different as in *cannabinus* the stem is more spongy, and the thorns fewer both in specimens from E. Indies and in cultivated ones and not beset below with several raised points; the leaves are broader, longer and smooth, the leafstalks much longer than the leaves and with very few thorns, and this also applies to the calyces. *H. radiatus* Willd. Sp. pl. 3. I p. 824 and Cavanilles Diss. 3 p. 150 Tab. 54, fig. 2 perhaps belongs here; but as Cavanilles says: "quae ante perfectam florum expansionem periit", he has described an incomplete plant and the comparison with our plant therefore remains uncertain S."

"Its stem is very tough, perhaps a kind of hemp could be prepared from this one and similarly from some other species of the same genus, but the natives do not make use of it. Th."

H. obtusatus Schum. 1827: 320 and 1829: 94 (no MS.).

Exsicc.: "Guinea", *Thonning* s.n. (**C**—2 sheets, type microf. 53: I. 7, II. 1–2).

"I leave it to others to determine whether this is a distinct species or merely a variety of the preceding, or a younger plant of *H. surattensis* S."

Note: Keay determined these specimens as *H. cannabinus* for FWTA. ed 2.— F.N.H.

Hibiscus micranthus *L. f.*—FTA. 1: 206 (1868); FWTA. ed. 2, 1: 346 (1958).

H. versicolor Schum. et Thonn. in Schum. 1827: 311 and 1829: 85; Junghans 1961: 342.

Exsicc.: "Guinea", *Thonning* 88 (**C**—1 sheet, type, microf. 54: I. 7; II. 1; **S**—1 sheet, type); "e Guinea" (**FI**—1 sheet, No. 25796).

"Here and there, flowers in May."
"Much resembles *H. rigidus* but the latter has oblong dentate leaves and the collar of the corolla is reflexed. S."

Sida alba sensu Junghans 1961: 343, pro parte non L.

Exsicc.: "Guinea", *Thonning* "9" (**C**—1 sheet, microf. 100: III. 5, 6).

Malvaceae indet. in Junghans 1961: 343.

Exsicc.: "Guinea in arenosis limosis", *Isert* s.n. (**C**—1 sheet, microf. 52: III. 3).

Hibiscus owariensis *P. Beauv.*—FWTA. ed. 2, 1: 347 (1958).

H. triumfettaefolius Thonning in Schum. 1827: 312 and 1829: 86; Junghans 1961: 342.

Exsicc.: Asiama, *Thonning* 302 (**C**—4 sheets, types, microf. 53: III. 6, 7; 54: I. 1–6; **S**—1 sheet, type).

"Grows near Asiama. Flowers in May."

Hibiscus surattensis *L.*—Schum. 1827: 317 and 1829: 91, 95; FWTA. ed. 2, 1 : 346 (1958); Junghans 1961 : 342.

Exsicc.: Aquapim, *Thonning* 197 (**C**—3 sheets, microf. 53: III. 1–5); "e Guinea" (**FI**—1 sheet, No. 25876; **P–JU**—1 sheet, No. 12372, microf. 916: I. 6).

Vernacular name: 'Sirsa-imum'.

"In damp places near Aquapim, not common."
"In the Indian plant the leaves are smaller and narrower than in the Guinea one and the leaves of the outer calyx have smaller appendages. In Cavanilles' figure the shape of the stipules is quite incorrectly represented; those in the Indian and Guinea plants are identical. S."

Hibiscus tiliaceus *L.*—Schum. 1827: 313 and 1829: 87 (no MS.); FWTA. ed. 2, 1 : 345 (1958); Junghans 1961 : 342.

Exsicc.: Volta R., *Thonning* s.n. (no material traced).

"Frequent on the sides of the Volta."

Hibiscus vitifolius *L.*—FTA. 1 : 197 (1868); FWTA. ed. 2, 1 : 346 (1958).

H. strigosus Schum. et Thonn. in Schum. 1827: 314 and 1829: 88; Junghans 1961 : 342.

Exsicc.: "Guinea", *Thonning* 18 (**C**—3 sheets, types, microf. 53: II. 3–7; **S**—1 sheet, type).

"Here and there, not rare."
"Comes nearest to *H. vitifolius* but the stem in the latter is less rough; the leaves shaggy and without the light streaks below; the calyx felted—shaggy especially before the flowers are unfolded, the outer calyx 7–8 leaved. S."

Hibiscus sp.

Malvaceae indet. in Junghans 1961 : 343.

Exsicc.: Sawi, 1785, *Isert* s.n. (**C**—1 sheet, microf. 52: III. 2).

Sida acuta *Burm. f.*—FWTA. ed. 2, 1 : 339 (1958). *S. carpinifolia* sensu FTA. 1 : 180 (1868), non L.

S. rugosa Thonning in Schum. 1827: 304 and 1829: 78; Junghans 1961 : 343.

Exsicc.: Aquapim, *Thonning* 260 (**C**—3 sheets, type, microf. 100: II. 2–7; **S**—1 sheet, type); "Guinea, *Horneman*" s.n. (**C**—1 sheet, microf. 100: I. 7; II. 1).

"At Aquapim."
"*S. rhombifolia* comes near it but differs in having obtuse leaves, blue-green underneath with shorter stalks; and in that the flower-stalks are much longer than the leaf-stalks. S."

Sida alba *L.*—FWTA. ed. 2, 1 : 339 (1958). *S. spinosa* sensu FTA. 1 : 180 (1868), non L.

S. scabra Thonning in Schum. 1827: 305 and 1829: 79; Junghans 1961: 343.

EXSICC.: "Guinea", *Thonning* 350 (**C**—3 sheets, microf. 100: III. 1–4; 101: I. 1, 2; **S**—1 sheet, type).

"Here and there, not very common."

"From the foregoing (*S. rugosa*) it differs especially in that the stem and leaf-stalks are often hairy; the leaves obtuse and although broad are nearer being ovate than lanceolate, blue-green underneath from an almost imperceptible felt; the stipules are bristle shaped; capsules are 5 in number, downy-haired but not 6–7 wrinkled and smooth. Th."

"In the cited characters it differs adequately from the related species *S. canescens*. S."

NOTE: One of the 4 specimens cited by Junghans under this species is *Hibiscus micranthus*—F.N.H.

Sida cordifolia *L.*—FTA. 1: 181 (1868); FWTA. ed. 2, 1: 339 (1958).

EXSICC.: Ada, 1784, *Isert* s.n. (**C**—1 sheet, microf. 101: I. 3).

S. decagyna Schum. et Thonn. in Schum. 1827: 307 and 1829: 81; Junghans 1961: 342.

EXSICC.: *Thonning* 119 (no type specimen traced).

"Grows here and there. Flowers in and after the rainy season."

Sida linifolia *Juss. ex Cav.*—FTA. 1: 179 (1868); FWTA. ed. 2, 1: 339 (1958).

S. linearifolia Thonning in Schum. 1827: 303 and 1829: 77; Junghans 1961: 343.

EXSICC.: "Guinea" coast, *Thonning* 120 (**C**—4 sheets, type, microf. 99: II. 3–7, III. 1–3); "e Guinea" (**FI**—1 sheet, No. 24930; **P–JU**—1 sheet, No. 12244, microf. 907: II. 2).

"Not very common in coastal areas. Flowers in June."

"*S. linifolia* differs in having 5 capsules without thorns. In *S. graminifolia* Rich. the leaves are broader and shorter, the racemes in the axils 2–3 flowered. S."

Urena lobata *L.*—FTA. 1: 189 (1868); FWTA. ed. 2, 1: 341 (1958).

U. diversifolia Schum. et Thonn. in Schum. 1827: 308 and 1829: 82; Junghans 1961: 343.

EXSICC.: Ada, 1784, *Isert* s.n. (**C**—1 sheet, microf. 108: II. 5); Quitta, *Thonning* 13 (**C**—2 sheets, type, microf. 108. II. 6, 7; III. 1, 2).

"Grows here and there at Ada and Quitta."

"Adequately differs from *U. reticulata* as in the latter the lowest leaves are trilobed and not palmate, the lobes are more obtuse, the serrations fewer and are more like teeth. S."

Wissadula amplissima *R. E. Fries* var. **rostrata** (*Schum. et Thonn.*) *R. E. Fries*—FTA. 1 : 182 (1868); FWTA. ed. 2, 1 : 336 (1958).

Sida rostrata Schum. et Thonn. in Schum. 1827 : 306 and 1829 : 80; Junghans 1961 : 343.

EXSICC.: Quitta, *Thonning* 121 (**C**—5 sheets, type, microf. 99: III. 4–7; 100: I. 1–6; **S**—1 sheet).

"In neighbourhood of Quitta towards the South. Flowers in October."

"Resembles *S. periplocifolia*, but in the latter the leaves are cordate-lanceolate, narrower, without sinuations; capsules much smaller and not pointed, flower-stalks, especially the single ones in the leaf-axils, much shorter. S."

MELASTOMATACEAE

Dissotis rotundifolia (*Sm.*) *Triana*—FWTA. ed. 2, 1: 257 (1954). *D. rotundifolia* var. *prostrata* (Thonning) Jac.-Fél. in Adansonia, sér. 2, 11: 548 (1971). *D. prostrata* (Thonning) Benth.—FTA. 2 : 452 (1871).

Melastoma prostrata Thonning in Schum. 1827 : 220 and 1828 : 240; Junghans 1961 : 343.

EXSICC.: "Guinea", *Thonning* 285 (**C**—2 sheets, types, microf. 66: III. 2–4); "e Guinea" (**P–JU**—1 sheet, No. 14051, microf. 1033: III. 5).

"At Blegusso and Aquapim."

"*M. repens* has ovate curved smooth crenate leaves, which on the margins are beset with short adpressed hairs, on the lower side light-green, short stalked. The calyx is urn-shaped, with interjacent smaller lobes and bedecked with single hairs. S."

[**Melastomastrum capitatum** (*Vahl*) *A. & R. Fernandes*—FWTA. ed. 2, 1: 761 (1958). *Dissotis erecta* (Guill. et Perr.) Dandy—FWTA. ed. 2, 1: 259 (1954).

Melastoma capitata Vahl 1796: 45 ("Habitat in India occidentali. Schumacher, Professor Chirurgiae Hauniensis"); Junghans 1961 : 343.

EXSICC.: (see note), ex Hb. Schumacher ex Banks and Dryander s.n. (**C**—1 sheet, microf. 66: II. 6, 7).

NOTE: Vahl assumed that this West African specimen came from the West Indies. Since Banks and Dryander are mentioned it is unlikely to be an Isert or Thonning specimen and I suggest it is one of Smeathman's as there is a specimen of this species collected by him in Sierra Leone at the British Museum.—F.N.H.]

Tristemma hirsutum *P. Beauv.*—FWTA. ed. 2, 1 : 250 (1954).

Melastomataceae indet. in Junghans 1961 : 343.

EXSICC.: Whydah, 1784, *Isert* s.n. (**C**—1 sheet, microf. 66: III. 1).

Tristemma incompletum *R. Br.*—FWTA. ed. 2, 1 : 250 (1954).

Melastoma sessilis Schum. et Thonn. in Schum. 1827: 219 and 1828: 239.

EXSICC.: Aquapim, *Thonning* 220 (**C**—3 sheets, types, microf. 66: III. 5–7; 67: I. 1, 2); "e Guinea" (**P–JU**—1 sheet, No. 14083, microf. 1035: III. 1).

"Grows in and near Aquapim."
"*M. capitata* Vahl Ecl. I, p. 45 which comes nearest to it, differs in its sharper, stiff-haired angles of the stem; in the leaves which are lanceolate, on upper side dark olive-green, with adpressed hairs and not rough; on lower side yellowish-green, smooth and veins and margins only hairy; floral leaves lanceolate hairy on the back. S."

Tristemma littorale *Benth.*—FWTA. ed. 2, 1 : 250 (1954).

Melastomataceae indet. in Junghans 1961 : 343.

EXSICC.: Whydah, 1785, *Isert* s.n. (**C**—1 sheet, microf. 66: II. 5).

MELIACEAE

Melia azedarach *L.*—FTA. 1 : 332 (1868); FWTA. ed. 2, 1 : 709 (1958).

M. angustifolia Schum. et Thonn. in Schum. 1827: 214 and 1828: 234; Junghans 1961 : 343.

EXSICC.: Elmina, *Thonning* 59 (**C**—1 sheet, type, microf. 67: I. 3, 4).

"Cultivated".
"*M. azedarach* differs in having fewer pinnae and in the leaflets, which are ovate-lanceolate, larger, more curved at the base, broader, almost cordate, more deeply serrate almost incised, the outer ones are more lengthily acuminate. The same holds good for *M. sempervirens*, see Swartz. Obs. p. 171. S."
"The seed has come from Elmina, presumably brought there from the West Indies. The flowers have a pleasant odour like that of lilac. Among all trees this one suffers most from the attacks of *Loranthus umbellatus*. Th."

Trichilia monadelpha (*Thonning*) *de Wilde* in Acta Bot. Neerl. 14: 455 (1965), and in Rev. Trichilia 108 (1968).

Limonia monadelpha Thonning in Schum. 1827: 217 and 1828: 237; FWTA. ed. 2, 1 : 705 (1958); Junghans 1962: 88.

EXSICC.: Aquapim, *Thonning* 263 (**C**—1 sheet, type, microf. 64: III. 5, 6).

"Grows at Aquapim".

MENISPERMACEAE

Chasmanthera dependens *Hochst.*—FWTA. ed. 2, 1 : 75 (1954).

Sterculiaceae indet. in Junghans 1962: 92 (see also *Melochia melissifolia* var. *bracteosa*).

EXSICC.: Whydah, 1785, *Isert* s.n. (**C**—1 sheet, microf. 105: I. 7).

MIMOSACEAE

Acacia nilotica (*L.*) *Willd. ex Del.* subsp. **adstringens** (*Schum. et Thonn.*) *Roberty*—Brenan in Kew. Bull. 21: 481 (1968). *A. nilotica* var. *adansonii* (Guill. et Perr.) Kuntze—FWTA. ed. 2, 1: 500 (1958).

Mimosa adstringens Schum. et Thonn. in Schum. 1827: 327 and 1829: 101; Junghans 1961: 344.

Exsicc.: "Guinea", *Isert* s.n. (**P–JU**—1 sheet, No. 14478, microf. 1064: II. 7); "Guinea", *Thonning* 239 (**C**—2 sheets, types, microf. 67: III. 2–5).

"At Ningo."
"The fruit has, owing to the gum-resin which occurs between its membranes a rather bitter astringent taste. Th."

Acacia pentagona (*Schum. et Thonn.*) *Hook. f.*—Brenan in FTEA. Legum.-Mimos. 100 (1959). *A. pennata* sensu FWTA. ed. 2, 2: 500 (1958), non (L.) Willd.

Mimosa pentagona Schum. et Thonn. in Schum. 1827: 324 and 1829: 98; Junghans 1961: 344.

Exsicc.: Jadofa, *Thonning* 274 (**C**—1 sheet, type, microf. 68: III. 3); "e Guinea" (**C**—2 sheets, microf. 68: III. 4–7; **FI**—1 sheet; **P–JU**—1 sheet, No. 14505, microf. 1066: I. 7).

"At Jadofa. Flowers in April."

Mimosaceae indet. in Junghans 1961: 344.

Exsicc.: *Mortensen* s.n. (**C**—2 sheets, microf. 69: I. 5–7; II. 1).

Albizia adianthifolia (*Schumacher*) *W. F. Wight*—FWTA. ed. 2, 1: 502 (1958).

Mimosa adianthifolia Schumacher 1827: 322 and 1829: 96; Junghans 1961: 344.

Exsicc.: Bligusso, *Thonning* 273 (**C**—3 sheets, types, microf. 67: II. 4–7; III. 1; **P–JU**—1 sheet, No. 14472, microf. 1064: I. 7).

"Grows near Bligusso and flowers in April."
"The male flowers are much larger than the rest; the tube of the filaments is double as short, the lobes numerous, the anthers sterile, no pistil. As regards the leaves it much resembles *Acacia gujanensis* Willd. but the spikes are in this latter filiform and cylindric. S."

Albizia glaberrima (*Schum. et Thonn.*) *Benth.*—FWTA. ed. 2, 1: 502 (1958).

Mimosa glaberrima Schum. et Thonn. in Schum. 1827: 321 and 1829: 95; Junghans 1961: 344.

Exsicc.: Asiama to Jadofa, *Thonning* 272 (**C**—3 sheets), types, microf.

68: I. 7; II. 1–4); "e Guinea" (**P–JU**—1 sheet, No. 14438, microf. 1061: III. 1).

VERNACULAR NAME: 'Laedjo-tjo'.

"Here and there between Asiama and Jadofa. Flowers in April."
"The wood is used as fuel. Th."

Calliandra portoricensis (*Jacq.*) *Benth.*—FTA. 2: 354 (1871); FWTA. ed. 2, 1: 504 (1958).

Mimosa guineensis Schum. et Thonn. in Schum. 1827: 323 and 1829: 97; Junghans 1961: 344.

EXSICC.: Aquapim, *Thonning* 195 (**C**—3 sheets, types, microf. 68: II. 5–7; III. 1, 2); "e Guinea" (**P–JU**—1 sheet, No. 14469, microf. 1064: III. 4).

"Grows in thickets and in neighbourhood of Aquapim."

NOTE: One sheet is apparently erroneously numbered 95 instead of 195.—F.N.H.

Dichrostachys cinerea (*L.*) *Wight et Arn.*—Brenan in FTEA. Legum.-Mimos. 36 (1959). *D. glomerata* (Forsk.) Chiov.—FWTA. ed. 2, 1: 494 (1958).

Mimosa bicolor Schum. et Thonn. in Schum. 1827: 326 and 1829: 100; Junghans 1961: 344.

EXSICC.: "Guinea", *Thonning* 133 (**C**—2 sheets, types, microf. 67: III. 6, 7; 68: I. 1, 2).

VERNACULAR NAME: 'Kahn-tjo'; fruit in Accra dialect: 'Beseri'.

"Grows here and there on bare ground."
"It exhausts the soil so that where it grows no other plant can thrive, although it is not overshadowed or choked by it. Because of its creeping roots it is difficult to eradicate. In cotton plantation near Fredriksberg (sic) it has almost choked cotton trees which otherwise are content with poor soil. Th."

Mimosa pigra *L.*—FWTA. ed. 2, 1: 495 (1958).

Mimosa canescens Willd. 1806: 1038 ("Habitat in Guinea. *Isert*"); Junghans 1961: 344.

EXSICC.: "Guinea", *Isert* s.n. (**B**—1 sheet, No. 19083, type, microf. 1384: I. 5; **C**—2 sheets, types, microf. 68: I. 3–6).

M. procumbens Schum. et Thonn. in Schum. 1827: 324 and 1829: 98; Junghans 1961: 344.

EXSICC.: Christiansborg, *Thonning* 65 (**C**—2 sheets, types, microf. 69: I. 1–4).

"A few bushes of this plant were discovered near Christiansborg."
"Resembles *M. asperata* Linn. so much in appearance that it might easily be taken for it; but in the latter the thorns are scattered along the partial leaf-stalk between the individual pinnae, whereas in the former [*M. procumbens*]

only the lowest part has them, the upper is without thorns; in *asperata* the thorns on the branches are reflexed, in our species straight and horizontal; in the first-named the hairs on the lomentum are so dispersed that one easily sees how it becomes opened transversely; in our plant the lomentum is beset with very dense hairs so that no part of the valve remains bare. I find that Persoon in Syn. pl. 2. p. 266 rightly distinguishes *M. pigra* and *M. asperata* which Wildenow unites sp. pl. 4, p. 1035."

NOTE: In spite of Schumacher's convincing comments above, *M. asperata* and *M. procumbens* are not maintained nowadays.—F.N.H.

Mimosa pudica *L.*—FWTA. ed. 2, 1 : 495 (1958).

Mimosaceae indet. In Junghans 1961 : 344.

EXSICC.: Whydah, 1785, *Isert* s.n. (**C**—1 sheet, microf. 69: II. 2, 3).

Tetrapleura tetraptera (*Schum. et Thonn.*) *Taub.*—FWTA. ed. 2, 1: 493 (1958). *T. thonningii* Benth.—FTA. 2: 330 (1871).

Adenanthera tetraptera Schum. et Thonn. in Schum. 1827: 213 and 1828: 233; Junghans 1961 : 344.

EXSICC.: Aquapim, *Thonning* 90 (**C**—2 sheets, type, microf. 2: II. 7; III. 1–3).

VERNACULAR NAME: 'Pepraemese'.

"Grows in fertile areas at Aquapim, flowers in May."

MOLLUGINACEAE

Gisekia pharnaceoides *L.*—FWTA. ed. 2, 1 : 134 (1954) and 759 (1958).

EXSICC.: Fida, 1784, Isert s.n. (**C**—1 sheet, microf. 44: III. 5).

G. linearifolia Schum. et Thonn. in Schum. 1827: 167 and 1828: 187; Junghans 1961 : 344.

EXSICC.: "Guinea", *Thonning* 176 (**C**—3 sheets, types, 44: III. 6, 7; 45: I. 1, 2); "e Guinea" (**P-JU**—1 sheet, No. 13400, microf. 992: II. 2).

"Here and there but not frequent."

Mollugo cerviana (*L.*) *Seringe*—FWTA. ed. 2, 1: 135 (1954).

Pharnaceum cerviana L.—Schum. 1827: 162 and 1828: 182 (no MS.); Junghans 1961: 344.

EXSICC.: Augna, 1784, *Isert* s.n. (**C**—1 sheet mounted with Thonning and Welwitsch specimen, microf. 79: II. 1); "Guinea", *Thonning* s.n. (**C**—1 specimen mounted on previous sheet).

"Here and there."

Mollugo nudicaulis *Lam.*—FWTA. ed. 2, 1: 134 (1954).

Pharnaceum spathulatum Sw.—Schum. 1827: 164 and 1828: 184; Junghans 1961: 345.

Exsicc.: Whydah, 1785, *Isert* s.n. (**C**—1 sheet, microf. 79: II. 2); "Guinea", *Thonning* 192 (**C**—2 sheets, microf. 79: II. 3–5).

"Common in cultivated places."

Glinus oppositifolius (*L.*) *A. DC.*—FWTA. ed, 2, 1: 135 (1954).

Pharnaceum mollugo L.—Schum. 1827: 163 and 1828: 183; Junghans 1961: 345.

Exsicc.: Ada, 1784, *Isert* s.n. (**C**—2 sheets, microf. 79: I. 4–6).

"Not very common."

MORACEAE

Ficus exasperata *Vahl* 1805: 197 ("Habitat in Guinea. *Isert*"); FTA. 6, 2: 110 (1916); FWTA. ed. 2, 1: 605 (1958); Junghans 1961: 345.

Exsicc.: Fida, 1785, *Isert* s.n. (**C**—1 sheet, type, microf. 42: II. 4).

Ficus scabra Willd. 1801: 102, Pl. 2, non Forst. f. 1786; Willd. 1806a: 1152 ("Habitat in Guinea").

Exsicc.: "Guinea", *Isert* s.n. (**B**—1 sheet, No. 19312, type, microf. 1399: III. 5; **S**—2 sheets, types).

Note: The Stockholm sheets were sent by Willdenow—F.N.H.

Ficus lutea *Vahl* 1805: 185 ("Habitat in Guinea. *Thonning*"); Schum. 1827: 25 and 1828: 45; FTA. 6, 2: 215 (1916); Junghans 1961: 345.

Exsicc.: *Thonning* 324 (no type material traced).

"Is cultivated."

Note: This remains an imperfectly known species owing to the lack of the type specimen.—F.N.H.

Ficus ovata *Vahl* 1805: 185 ("Habitat in Guinea. *Thonning*"); Schum. 1827: 26 and 1828: 46; FTA. 6, 2: 164 (1916); FWTA. ed. 2, 1: 608 (1958); Junghans 1961: 345.

Exsicc.: Between Christiansborg and Fredericksberg, *Thonning* 246 (no type material at **C**; **S**—1 sheet, type).

Vernacular name: 'Ninndu-tjo'.

"On roads between Christiansborg and Frederiksberg."
"A tree growing very quickly to a moderate height. The better the soil is the more does it shoot straight up with almost perpendicular branches; on the other hand in very poor soil it spreads out laterally especially where the ground is so stony that the roots can only penetrate it with difficulty; on planting, it very readily emits roots."

F. calyptrata Vahl 1805: 186 ("Habitat in Guinea. *Thonning*"); Schum. 1827: 27 and 1828: 47.

Exsicc.: *Thonning* 329 (no type material traced).

Vernacular name: 'Apaataa'.

"Here and there."
"The fruit is eaten by the natives."

Ficus polita *Vahl* 1805: 182 ("Habitat in Guinea. *Isert*"); FWTA. ed. 2, 1: 611 (1958); Junghans 1961: 345.

Exsicc.: Whydah, 1785, *Isert* s.n. (C—1 sheet, type, microf. 42: II. 5).

Ficus thonningii *Blume*—FTA. 6, 2: 188 (1916); FWTA. ed. 2, 1: 610 (1958).

F. microcarpa Vahl 1805: 188 ("Habitat in Guinea. *Thonning*"); Schum. 1827: 28 and 1828: 48; Junghans 1961: 345, non L. f.

Exsicc.: *Thonning* 325 (no type material traced).

Ficus umbellata *Vahl* 1805: 182 ("Habitat in Guinea. *Thonning*"); Schum. 1827: 25 and 1828: 45; FWTA. ed. 2, 1: 610 (1958).

Exsicc.: *Thonning* 326 (no type material traced).

"Because of its quick growth and good shade it is planted commonly in the broadest streets and in the squares of native communities."

Musanga cecropioides *R. Br.*—FWTA. ed. 2, 1: 616 (1958).

Moraceae indet. in Junghans 1961: 345.

Exsicc.: Aquapim, 1786, *Isert* s.n. (C—1 sheet, microf. 69: II. 7; III. 1).

MYRTACEAE

Eugenia coronata *Schum. et Thonn.* in Schum. 1827: 230 and 1829: 4; FWTA. ed. 2, 1: 237 (1954); Junghans 1961: 346.

Exsicc.: "Guinea", 1785, *Isert* s.n. (C—1 sheet, microf. 40. II. 5); "Guinea" coast, *Thonning* 53 (C—1 sheet, type, microf. 40: III. 1); "e Guinea" (C—2 sheets, microf. 40: II. 6, 7); G–DC—1 sheet, 3: 271. 69, type, microf. 539: I. 8; P–JU—1 sheet, No. 13939, microf. 1026: III. 6).

Vernacular name: 'Amuma'.

"Frequent especially at the seaside."
"The natives eat the fruit. Near the shore it is scarcely an ell in height; but inland it is as large as *Prunus spinosa* with which it has at a distance much resemblance when in flower. Th."

Eugenia jambos *L.*—FWTA. ed. 2, 1: 238 (1954).

Myrtaceae indet. in Junghans 1961: 346.

Exsicc.: Fida, 1785, *Isert* s.n. (C—1 sheet, microf. 40: III. 2).

Psidium guajava *L.*—FWTA. ed. 2, 1 : 235 (1954).

P. longifolium Schumacher 1827: 229 and 1829: 3 (no MS.); Junghans 1961: 346.

Exsicc.: "Guinea", 1783, *Isert* s.n. (**C**—1 sheet, type, microf. 89: II. 6).

Syzygium guineense *(Willd.) DC.* var. **guineense**—FWTA. ed. 2, 1 : 240 (1954).

Calyptranthes guineensis Willd. 1800a: 974 ("Habitat in Guinea"); Junghans 1961: 346.

Exsicc.: Sawi, 1785, *Isert* s.n. (**B**—1 sheet, No. 9582, type, microf. 660: II. 6; **C**—1 sheet, type, microf. 13: III. 3).

Syzygium guineense *(Willd.) DC.* var. **littorale** *Keay*—FWTA. ed. 2, 1 : 240 (1954).

Myrtaceae indet. in Junghans 1961: 346.

Exsicc.: Fida, 1784, *Isert* s.n. (**C**—1 sheet, microf. 40: III. 3).

NYCTAGINACEAE

Boerhavia diffusa *L.*—Schum. 1827: 16 and 1828: 36; FWTA. ed. 2, 1 : 178 (1954); Junghans 1961: 346.

Exsicc.: "Guinea", *Thonning* 17 (**C**—1 sheet, microf. 11: I. 4).

Vernacular name: 'Tjalala'.

"Grows commonly near towns and in cultivated places."
"Swartz's description in Observat. p. 11, belongs rather to the following species" [*B. adscendens*].
"Natives boil the root bark in soup, which they drink for dysentery; moreover it is one of the commonest fetish plants, which are used by the natives for their cleansing baths in sicknesses or other cases. Th."

B. adscendens Willd. 1797a: 19 ("Habitat in Guinea"); Vahl 1804: 285 ("Habitat in Guinea. *Isert, Thonning.*"); Schum 1827: 17 and 1828: 37.

Exsicc.: "Guinea", *Isert* s.n. (**B**—1 sheet, No. 768, type, microf. 46: II. 6); "Guinea", *Thonning* 97 (**C**—2 sheets, types?, microf. 11: I. 2, 3, 5).

Vernacular name: 'Tjalala'.

"Habitat as in the preceding" [*B. diffusa*].
"Uses as in the preceding one. Th."

NYMPHAEACEAE

Nymphaea guineensis *Schum. et Thonn.* in Schum. 1827: 248 and 1829: 22; Junghans 1961: 346; Hepper in Kew Bull. 26: 566 (1972).

Exsicc.: Quitta, *Thonning* 149 (**C**—2 sheets, type, microf. 70: III. 2–5).

Vernacular name: 'Taetremande'.

"Not frequent. Near Quitta."

N. rufescens Guill. et Perr. 1831 : 15; FWTA. ed. 2, 1 : 66 (1954).

Nymphaea lotus *L.*—FWTA. ed. 2, 1 : 66 (1954).

N. dentata Schum. et Thonn. in Schum. 1827: 249 and 1829: 23; Junghans 1961 : 346.

Exsicc.: *Thonning* 183 (no type material traced).

Vernacular name: 'Taetremande'.

"In stagnant water. Flowers in July."
"More exact investigations will show whether this plant is different from-*N. lotus*; it is at least a remarkable sub-species. The leafstalk in *N. lotus* is just attached to the carved out angle, which the figures in Alpini Exot. p. 213 and 224 and Vesling Observ. p. 43 demonstrate; and not above the indentation. S."

Nymphaea maculata *Schum. et Thonn.* in Schum. 1827: 247 and 1829: 21; FWTA. ed. 2, 1 : 66 (1954); Junghans 1961 : 346.

Exsicc.: "Guinea", *Thonning* 175 (C—2 sheets, type, microf. 70: III. 6, 7; 71: I. 1, 2).

Vernacular name: 'Taetremande'.

"In stagnant water. Flowers in July."

OCHNACEAE

Ochna membranacea *Oliv.*—FWTA. ed. 2, 1 : 222 (1954).

Exsicc.: Aflahu, 1784, *Isert* s.n. (C—1 sheet, microf. 47: III. 1).

Ochna multiflora *DC.*—FWTA. ed. 2, 1 : 223 (1954).

Exsicc.: Popo, 1785, *Isert* s.n. (C—1 sheet, microf. 47: II. 7); "e Guinea", *Thonning* s.n. (C—1 sheet, microf. 47: III. 3).

Ouratea flava *(Schum. et Thonn.) Hutch. et Dalz. ex Stapf*—FWTA. ed. 2, 1 : 228 (1954). *Gomphia reticulata* sensu FTA. 1: 320 (1868), partly. *Campylospermum flavum* (Schum. et Thonn.) Farron in Bull. Jard. Bot. Brux. 35: 397 (1965).

Gomphia flava Schum. et Thonn. in Schum. 1827: 216 and 1828: 236; Junghans 1961 : 346.

Exsicc.: "Guinea", *Thonning* 72 (C—2 sheets, type, microf. 47: II. 3–6); "e Guinea" (C—1 sheet, microf. nil; P–JU—1 sheet, No. 13018 [5], microf. 961: III. 4).

"Grows in the fertile mountain regions, and flowers at different times of the year."

Ochnaceae indet. in Junghans 1961 : 346.

EXSICC.: Aquapim, 1786, *Isert* s.n. (**C**—1 sheet, microf. nil).

OLACACEAE

Ximenia americana *L.*—Schum. 1827: 193 and 1828: 213; FWTA. ed. 2, 1 : 646 (1958); Junghans 1961: 347.

EXSICC.: Christiansborg, *Thonning* 91 (**C**—1 sheet, microf. 111: II. 3, 4); "e Guinea" (**P–JU**—1 sheet, No. 11893, microf. 882: I. 6).

VERNACULAR NAME: 'Me-tio'.

"Not common; a mile north of Christiansborg. Flowers in June."

OLEACEAE

Jasminum dichotomum *Vahl* 1804: 26 ("Habitat in Guinea. *Thonning*"); Schum. 1827: 7 and 1828: 27; FWTA. ed. 2, 2: 50 (1963); Junghans 1961: 347.

EXSICC.: "Guinea", *Thonning* 76 (**C**—2 sheets, types, microf. 61: II. 2–5); "e Guinea"? (**C**—1 sheet, microf. 61: II. 6).

VERNACULAR NAME: 'Jangkumaetri'.

"Common amongst other bushes; it flowers chiefly at night and exhales an agreeable scent. The size of flower and the scent are as in *J. officinale*."

"The natives crush the leaves and apply them in that condition to old bone-injuries, after the wound has been cleansed by other means. Th."

ONAGRACEAE

Ludwigia octovalvis (*Jacq.*) *Raven* subsp. **brevisepala** (*Brenan*) *Raven* in Reinwardtia 6: 365 (1963). *Jussiaea suffruticosa* L. var. *linearis* (Willd.) Oliv. ex Kuntze—FWTA. ed. 2, 1 : 169 (1954).

Jussieua linearis Willd. 1799: 575 ("Habitat in Guinea"); Schum. 1827: 217 and 1828: 237; Junghans 1961 : 347.

EXSICC. Whydah, 1784, *Isert* s.n. (**B**—1 sheet, No. 8131, type, microf. 555: III. 2; **C**—1 sheet, type, microf. 61: III. 4); "Guinea", *Thonning* 82 (**C**—4 sheets, types, 61: III. 5–7; 62: 1, 2); "e Guinea" (**FI**—1 sheet; **P–JU**—1 sheet, No. 13745, microf. 1015: I. 4).

"Grows here and there. Flowers now and then after continuous rain."

Ludwigia abyssinica *A. Rich.*—Raven in Reinwardtia 6: 380 (1963). *Jussiaea abyssinica* (A. Rich.) Dandy et Brenan—FWTA. ed. 2, 1 : 170 (1954).

Onagraceae indet. in Junghans 1961 : 347.

EXSICC.: Whydah, 1785, *Isert* s.n. (**C**—1 sheet, microf. 72: III. 3).

PAPAVERACEAE

Argemone mexicana *L.*—Schum. 1827: 233 and 1829: 7; FWTA. ed. 2, 1: 84 (1954); Junghans 1961 : 347.

Exsicc.: Akkra, *Thonning* 157 (**C**—1 sheet, microf. 6: II. 6).

"Grows at Akkra."
"Plant contains a yellow sap."

PAPILIONACEAE

Abrus precatorius *L.* subsp. **africanus** *Verdc.* in Kew Bull. 24: 241 (1970).

A. precatorius L.—Schum. 1827: 332 and 1829: 106 (no MS. No.); FWTA. ed. 2, 1 : 574 (1958); Junghans 1961 : 347.

Exsicc.: "Guinea", *Thonning* s.n. (**C**—1 sheet, microf. nil).

NOTE: This sheet has been found since Junghans wrote his paper in which he recorded the species without having traced any material—F.N.H.

Aeschynomene indica *L.*—FTA. 2: 147 (1871); FWTA. ed. 2, 1: 580 (1958).

A. quadrata Schum. et Thonn. in Schum. 1827: 356 and 1829: 130; Junghans 1961 : 347.

Exsicc.: Ga, Adampi, Augna, *Thonning* 147 (**C**—3 sheets, types, microf. 3: II. 3–6, only 2 sheets photographed).

"In Ga, Adampi, Augna; not common. Found in flower in Oct., Nov."
"It differs sufficiently from *A. diffusa* in its stiff-haired stem, branch and flower-stalk, in its more and larger leaves which are devoid of veins on lower surface; and in the entire habit of the plant. S."

Aeschynomene sensitiva *Sw.*—FWTA. ed. 2, 1 : 580 (1958).

Papilionaceae indet. of Junghans 1961 : 355.

Exsicc.: Sawi, *Isert* s.n. (**C**—1 sheet, microf. 3: II. 2).

Alysicarpus ovalifolius (*Schum. et Thonn.*) *J. Léonard* in Bull. Jard. Bot. Brux. 24: 88, fig. 11 (1954); FWTA. ed. 2, 1 : 587 (1958); Schubert 1963: 295.

Hedysarum ovalifolium Schum. et Thonn. in Schum. 1827: 359 and 1829: 133 (no MS No.); Junghans 1961 : 351.

Exsicc.: Ada, 1784, *Isert* s.n. (**C**—2 sheets, microf. 51: II. 7; III. 1–3); Ada, *Thonning* s.n. (**C**—2 sheets, types, microf. 51: II. 4–6); "e Guinea" (**P–JU**—1 sheet, No. 15498, microf. 1139: III. 4).

"At Ada."

"Differs from *H. moniliforma* in having more shortly stalked leaves, and stipules of same length as the leaf-stalks; from *H. vaginale* in the downy haired lomenta of the racemes. S."

Alysicarpus rugosus *(Willd.)* *DC.*—FTA. 2: 171 (1871); J. Léonard in Bull. Jard. Bot. Brux. 24: 92, fig. 12 (1954); FWTA. ed. 2, 1: 587 (1958); Schubert 1963: 296.

Hedysarum rugosum Willd. 1802: 1172 ("Habitat in Guinea"); Schum. 1827: 358 and 1829: 132; Junghans 1961: 351.

Exsicc.: "Guinea", *Isert* s.n. (**B**—1 sheet, No. 13757, type, microf. 983: III. 8); "Guinea", *Thonning* 188 (no material at **C**); "e Guinea" (**FI**—1 sheet; **P–JU**—1 sheet, No. 15575, microf. 1145: III. 6).

"Not common. Flowers in June."
"It agrees with *H. glumaceum* in the 4-partite calyx and in that the segments of the lomentum are transversely wrinkled; but it differs in the hairy stem and leaf; has longer and more obtuse leaves, more and larger flowers, leaves near the flowers and shorter lomenta. S."

Arachis hypogaea *L.*—Schum. 1827: 337 and 1829: 111; FWTA. ed. 2, 1: 576 (1958); Junghans 1961: 347.

Exsicc.: "Guinea", 1786, *Isert* s.n. (**C**—1 sheet, microf. 6: II. 5); "Guinea", *Thonning* 124 (**C**—1 sheet, microf. 6: II. 3, 4, also 1 sheet previously indet. not photographed).

Vernacular names: "Engkatje" (Ashanti); "Molaque" (Accra); "Assianthé bönner" (Danish).

"It is pretty commonly cultivated but in slight quantities; it thrives nearly everywhere and yields a rich harvest."
"The seeds which are very rich in a thick and mild oil, are eaten by the natives; the husk is roasted till it falls off."

Atylosia scarabaeoides *(L.)* *Benth.*—FWTA. ed. 2, 1: 560 (1958); Hepper in Kew Bull. 28: 319 (1973).

Glycine mollis Willd. 1802: 1062 ("Habitat in Guinea?"); Junghans 1961: 350.

Exsicc.: "Guinea", *Isert* s.n. (**B**—1 sheet, No. 13446, type, microf. 959: I. 9).

Baphia nitida *Lodd.*—FTA. 2: 249 (1871); FWTA. ed. 2, 1: 512 (1958).

Podalyria haematoxylon Thonning in Schum. 1827: 202 and 1828: 222; Junghans 1961: 354.

Exsicc.: Aquapim, *Thonning* 48 (**C**—3 sheets, types, microf. 87: II. 4–7, III. 1, 2); "e Guinea" (**P–JU**—1 sheet, No. 14701, microf. 1080: II. 1).

"Grows most often in the valleys at Aquapim, although it is not at all common in this place."
"The wood has a light colour, sometimes mixed with reddish veins; it is fairly fine but not to be obtained in large pieces and not fine enough for

small work. After the tree has died the wood gets a much darker red colour; it dies out and putrefies from inside outwards, so that at last only the external part is left which is generally so ravaged by worms and putrescence that it is unserviceable for working. For dyeing on the contrary it is most excellent as it is exceedingly rich in colouring matter. It is much used by the natives for fetish ceremonies and amulets; for this purpose it is finely macerated with water on a stone. Th."

Cajanus cajan (*L.*) *Millsp.*—FTA. 2: 216 (1871); FWTA. ed. 2, 1: 559 (1958).

Cytisus guineensis Schumacher 1827: 349 and 1829: 123 ("Ved Whyda efter Isert"); Junghans 1961: 348.

Exsicc.: Whydah, 1785, *Isert* s.n. (**C**—1 sheet, type, microf. 35: I. 1, 2); "Guinea", *Thonning* s.n. (**C**—1 sheet, microf. 34: III. 7).

"Near Whyda according to Isert."

"Agrees with *C. cajan* as regards the shape of the leaf, but differs from it in that in *C. cajan* the leaves are white with silky hairs above, in *C. guineensis* dark-green; in former the flower-stalks are nearly double as long as the leaves, in latter the length of the leaf-stalks. S."

"Cultivated by inhabitants although rarely. Th."

Canavalia rosea (*Sw.*) *DC.*—FWTA. ed. 2, 1: 574 (1958).

Dolichos obovatus Schum. et Thonn. in Schum. 1827: 341 and 1829: 115; Junghans 1961: 349.

Exsicc.: "Guinea" coast, *Thonning* 64 (**C**—3 sheets, microf. 37: I. 6, 7; II. 1–3); "e Guinea" (**P–JU**—1 sheet, No. 15150, microf. 1113: II. 1).

Vernacular name: 'Ammba-pang'.

"Rather uncommon among the shrubs along the coast, not occurring further inland."

"*D. rotundifolius* Vahl, Symb. Bot. 2. p. 81 comes very near it, but has downy-haired pods. S."

"The natives make use at times of the stem for tying. Th."

Canavalia virosa (*Roxb.*) *Wight et Arn.*—Sauer in Brittonia 16: 152 (1964).

Dolichos ovalifolius Schum. et Thonn. in Schum. 1827: 342 and 1829: 116; Junghans 1961: 349.

Exsicc.: Dudua, *Thonning* 221 (**C**—1 sheet, type, microf. 37: II. 7); "e Guinea" (**P–JU**—1 sheet, No. 15157, microf. 1113: III. 1).

"At Dudua."

Papilionaceae indet. in Junghans 1961: 355.

Exsicc.: Whydah, 1785, *Isert* s.n. (**C**—1 sheet, microf. 37: III. 4, 5).

Crotalaria glauca *Willd.* 1802: 974 ("Habitat in Guinea"); DC. 1825a: 127; Schum. 1827: 334 and 1829: 108; FTA. 2: 12 (1871); FWTA. ed. 2, 1: 548 (1958); Junghans 1961: 348.

Exsicc.: "Guinea," *Isert* s.n. (**B**—1 sheet, No. 13242, type, microf. 945: II. 6); "Guinea", *Thonning* 182 (**C**—2 sheets, types, microf. 27: III. 4–6; **S**—1 sheet, type); "e Guinea" (**P–JU**—1 sheet, No. 14907, microf. 1093: III. 3?).

"Here and there in open fields amongst the young grass. Flowers in June and Sept."

"*C. graminea* Herb. Vahl much resembles our plant but differs in its linear-lanceolate more pointed leaves; it is from East Indies. S."*

C. genistifolia Vahl ex Schumacher in Schum. 1827: 335 and 1829: 109 (no MS); Junghans 1961: 348.

Exsicc.: "e Guinea", *Isert* s.n. (**C**—2 sheets, types, microf. 27: III. 2, 3; **S**—1 sheet, type).

"This plant was found by Isert in Guinea and communicated to v. Rohr who in his turn passed it on to our herbaria. It is certainly different from the foregoing species [*C. glauca*]; in *C. glauca* the stem is branched below, in *genistifolia* only above and almost dichotomously; the former has erect nearly adpressed leaves, in the latter they are divergent (patentia); in the former the 3 or several racemes are lateral with 1–2 or 4 flowers, nearly 2 inches long, and as long as the leaves; in the latter only one raceme which is erect 6–8 flowered, attached to the tip of the branch and only slightly longer (about 4 inches long). S."

Crotalaria goreensis *Guill. et Perr.*—FWTA. ed. 2, 1: 548 (1958); Polhill in Kew Bull. 22: 190 (1968).

C. falcata Schum. et Thonn. in Schum. 1827: 335 and 1829: 109, non Vahl ex DC., *nom illegit.*; Junghans 1961: 348, partly.

Exsicc.: "Guineae", 1786, *Isert* s.n. (**C**—1 sheet, microf. 27: II. 7); Akra, Adampi and Augna, *Thonning* 189 (**C**—4 sheets, types, microf. 27: II. 1–6; III. 1); "e Guinea" (**FI**—1 sheet; **P–JU**—1 sheet, No. 14924B, microf. 1094: III. 3).

"Rather frequent in Akra, Adampi, Augna".

Crotalaria pallida *Ait.*

C. pallida *Ait.* var. **pallida**—Polhill in Kew Bull. 22: 262 (1968).

C. striata DC. 1825a: 131; Schum. et Thonn. 1827: 336 and 1829: 110; Junghans 1961: 348.

Exsicc.: Quitta, *Thonning* 40 (**C**—2 sheets, types, microf. 28: II. 4–6).

"Here and there, e.g. at Quitta".

C. pallida *Ait.* var. **obovata** (*G. Don*) *Polhill* l. c. 265 (1968).

C. falcata Vahl ex DC. 1825a: 132; FWTA. ed. 2, 1: 551 (1958); Junghans 1961: 348 partly (excluding *Thonning* 189 which is *C. goreensis*).

* Two König specimens from India are included in the Vahl herbarium and shown in the microfiche No. 2203 on card 28: I. 1–4.

Exsicc.: "Guinea", *Thonning* s.n. (**G–DC**—1 sheet, No. 2: 132. 95, type, microf. 308: II. 4; **S**—1 sheet, type).

C. obovata G. Don 1832: 138 ("Native of Guinea"); Junghans 1961: 348.

Exsicc.: "e Guinea misit Dr. Von Rohr" (**C**—3 sheets, microf. 28: I. 5–7; II. 1–3; **P–JU**—1 sheet, No. 14925, microf. 1094: III. 4).

Dalbergia ecastaphyllum (*L.*) *Taub.*—FWTA. ed. 2, 1: 515 (1958).

Ecastaphyllum brownei Pers.—Schum. 1827: 332 and 1829: 106; Junghans 1961: 349.

Exsicc.: Volta R., *Thonning* 319 (no material traced).

"Common on banks of the Volta."

Desmodium gangeticum (*L.*) *DC.*—FWTA. ed. 2, 1: 584 (1958); Schubert 1963: 294.

Hedysarum lanceolatum Schum. et Thonn. in Schum. 1827: 360 and 1829: 134; Junghans 1961: 351.

Exsicc.: Aquapim, *Thonning* 201 (**C**—2 sheets, types, microf. 51: II. 1–3); "e Guinea" (**P–JU**—1 sheet, No. 15577, microf. 1146: I. 1).

"In thickets in the valleys near Aquapim. Flowers June-August."
"Very closely related to *H. gangeticum*, but differs in having ovate-lanceolate leaves; of which the upper are narrower; the branches form moreover sharper angles which are more hairy; and the lomenta are beset with more and longer hooks. S."

Desmodium ramosissimum *G. Don.*—FWTA. ed, 2, 1: 584 (1958); Schubert 1963: 293. *D. mauritianum* (Willd.) DC.—FTA. 2: 164 (1871).

Hedysarum fruticulosum Schum. et Thonn. in Schum. 1827: 363 and 1829: 137; Junghans 1961: 350, non *H. fruticulosum* Desv. (1826).

Exsicc.: Whydah, 1785, *Isert* s.n. (**C**—1 sheet, microf. 51: I. 1, 2); Aquapim, *Thonning* 203 (**C**—2 sheets, types, microf. 50: III. 5–7); "e Guinea" (**FI**—1 sheet; **P–JU**—1 sheet, No. 15580, microf. 1146: I. 4).

Vernacular name: 'Alipoma-kirpei'.

"In neighbourhood of Aquapim, not very common."

Desmodium triflorum (*L.*) *DC.*—FTA. 2: 166 (1871); FWTA. ed. 2, 1: 584 (1958); Schubert 1963: 293.

Hedysarum granulatum Schum. et Thonn. in Schum. 1827: 362 and 1829: 136; Junghans 1961: 351.

Exsicc.: Accra, 1784, *Isert* s.n. (**C**—2 sheets, microf. 51: I. 3, 6, 7); "Guinea," *Thonning* 187 (**C**—1 sheet, type, microf. 51: I. 4, 5); "e Guinea" (**FI**—1 sheet; **P–JU**—1 sheet, No. 15576, microf. 1145: III. 7).

"Not common, grows and flowers in June and July."

"Approaches *H. triflorum* but differs in that the racemes are situated in the leaf-axils. From *H. repens* it differs in the obcordate emarginate leaflets. S."

Hedysarum granuliferum Sprengel, Mant. Prima Fl. Halensis 48 (1817).

EXSICC.: "e Guinea", *Thonning* (type not traced).

NOTE: The type probably went to Berlin with residue of Sprengel's herbarium and now is lost. However, it must have been a duplicate of *Thonning* 187 cited above.—F.N.H.

Desmodium velutinum (*Willd.*) *DC.*—FWTA. ed. 2, 1: 584 (1958); Schubert 1963: 292. *D. lasiocarpum* (P. Beauv.) DC.—FTA. 2: 162 (1871).

Hedysarum deltoideum Schum. et Thonn. in Schum. 1827: 361 and 1829: 135; Junghans 1961: 350.

EXSICC.: Whydah, 1785, *Isert* s.n. (**C**—2 sheets, as *H. umbrosum*, microf. 52: I. 3, 4); Aquapim, *Thonning* 202 (**C**—2 sheets, types, microf. 50: III. 1–4); "e Guinea" (**FI**—1 sheet; **P–JU**—1 sheet, No. 15578, microf. 1146: I. 2).

"Amongst shrubs in formerly cultivated places in valleys near Aquapim. Flowers in June and July."
"Approaches *H. velutinum* but as the description indicates, sufficiently differs from it. There occurs in Isert's herbarium a plant under the name of *H. umbrosum* with more pointed leaves, less striped, less felted and reddish-brown branches; perhaps a variety of the foregoing, perhaps a distinct species; it was found by Isert near Whydah. S."

Papilionaceae indet. in Junghans 1961: 355.

EXSICC.: "e Guinea" (**C**—1 sheet, microf. 52: I. 1, 2).

Dioclea reflexa *Hook. f.*—FWTA. ed. 2, 1: 574 (1958).

Papilionaceae indet. in Junghans 1961: 355.

EXSICC.: Whydah, 1785, *Isert* s.n. (**C**—1 sheet, microf. 37: III. 1); "Guinea", *Thonning* s.n. (**C**—1 sheet, microf. 37: II. 6).

Drepanocarpus lunatus (*L. f.*) *G. F. W. Mey.*—FTA. 2: 237 (1871); FWTA. ed. 2, 1: 519 (1958). *Machaerium lunatum* (L. f.) Ducke in Arch. Jard. Bot. Rio de Janeiro 4: 310 (1925).

Sommerfeldtia obovata Schum. et Thonn. in Schum. 1827: 331 and 1829: 105; Bentham in Journ. Linn. Soc. Bot. 4 (Suppl.): 69 (1860); Junghans 1961: 354.

EXSICC.: Ada, 1784, *Isert* s.n. (**C**—1 sheet, microf. 103: III. 2, 3); Volta banks, *Thonning* 254 (**C**—2 sheets, types, microf. 103: II. 6, 7; III. 1).

VERNACULAR NAME: 'Ohoa-tjo'.

"Common on banks of the Volta."
"I have named this genus after the Norwegian botanist Pastor Christian Sommerfeldt, well known for his Supplement to Wahlenberg's Flora Lapponica and several botanical works. S."

Eriosema laurentii *De Wild.*—FTEA. Legum.—Papil. 771 (1971). *E. glomeratum* (Guill. et Perr.) Hook. f.—FTA. 2: 228 (1871); FWTA. ed. 2, 1: 558 (1958). *Eriosema rufum* (Schum. & Thonn.) Baill. in Adansonia 6: 226 (1866), non (Kunth) G. Don, *nom. illeg.*

Glycine rufa Thonning in Schum. 1827: 344 and 1829: 118 (no MS No.); Junghans 1961: 350, non Kunth.

Exsicc.: Quitta, *Thonning* s.n. (**C**—3 sheets, types, microf. 46: II. 5–7; III. 1, 2); "e Guinea" (**P–JU**—1 sheet, No. 15177, microf. 1116: I. 6).

"Frequent near Quitta."
"It agrees closely with *Crotalaria picta* Vahl, Symbol. 2 p. 81 as regards shape of the leaf and the reddish-brown appearance of the whole plant, so that without the inflorescence it could be easily taken for it; but *C. picta* has racemes in the leaf-axils and flower-stalks of about 1½ inches in length; in *Glycine rufa* on the other hand the flowers in the leaf-axils are aggregated into heads and the flower-stalks are not ½ inch long. S."

Erythrina senegalensis *DC.*—FTA. 2: 181 (1871); FWTA. ed. 2, 1: 562 (1958).

E. latifolia Schum. et Thonn. in Schum. 1827: 333 and 1829: 107.

Exsicc.: "Guinea", *Thonning* 104 (**C**—1 sheet, type, microf. 40: I. 6–7).

Vernacular name: 'Naba-tiölu'.

"Occurs rarely wild but is cultivated by the natives. A decoction of the bark is used by the natives against dysentery and colic and for easing difficult deliveries."

Galactia tenuiflora *(Willd.) Wight et Arn.*—FWTA. ed. 2, 1: 563 (1958). *Glycine tenuiflora* Willd. 1803: 1059.

Papilionaceae indet. in Junghans 1961: 355.

Exsicc.: Ada, 1784, *Isert* s.n. (**C**—1 sheet, microf. 46: III. 3, 4).

[**Galactia rubra** *(Jacq.) Urb.* in Symb. Antill. 2: 309 (1900).

Glycine sericea Willd. 1802: 1059 ("Habitat in Guinea"); DC. 1825a: 242; Junghans 1961: 350.

Exsicc.: Martinique, *Isert* s.n. (**B**—1 sheet, No. 13432, type, microf. 958: III. 1).

Note: Although Willdenow attributes this to Isert from Guinea, the species is West Indian; Urban (l.c. 310) considers it was collected in Martinique, and it could have been collected by Isert during his return voyage—F.N.H.]

Indigofera aspera *Perr. ex DC.*—FWTA. ed. 2, 1: 538 (1958). *I. senegalensis* Lam.—FTA. 2: 102 (1871).

I. tenella Schumacher 1827: 367 and 1829: 141; Junghans 1961: 353.

Exsicc.: Quitta (fl. & fr. Nov.) *Thonning* 127 (**C**—1 sheet, type, microf. 57: III. 5, 6).

"Only found at Quitta in loose sandy ground; in Nov. with flower and seeds."

"Very like *I. senegalensis* Lamarck Encyclopedia but yet different from it. In the latter the leaves are broader and are placed 7 together, the racemes are three times as long as the leaves, the pods longer and more sickle-shaped; in *I. tenella* on the other hand the leaves are linear and placed 5 together, the racemes much shorter, the pods shorter less sickle-shaped and compressed. S."

"It is too insignificant in growth to be cultivated usefully. Th."

Indigofera colutea (*Burm. f.*) *Merrill*—FWTA. ed. 2, 1 : 540 (1958).

I. hirsuta sensu Jacq. 1788: 359–60 ("Crescit in Guinea and Zeylona") as to plant from Guinea; 1786–93b: pl. 569; Junghans 1961: 352.

Exsicc.: probably *Isert* s.n. (no material traced).

Indigofera dendroides *Jacq.* 1788: 357–8 ("Crescit in arenosis Guineae"); 1786–93b: pl. 571; Willd. 1802: 1235; Schum. 1827: 375 and 1829: 149; FWTA. ed. 2, 1 : 540 (1958); Junghans 1961 : 351.

Exsicc.: Christiansborg, *Isert* s.n. (**B**—1 sheet, No. 13903, type, microf. 994: I. 4; **C**—1 sheet, type, microf. 59: I. 7; II. 1); "Guinea", *Thonning* 36 (**C**—6 sheets, microf. 59: I. 3–6; II. 2–7); "e Guinea" (**FI**—1 sheet; **G–DC**—1 sheet, 2: 227. 59, microf. 349: III. 4; **P–JU**—1 sheet No. 15382, microf. 1130: III. 3).

"Here and there especially on cultivated areas."

"With the lens one sees that the leaves are beset with adpressed hairs. S."

"Its appearance seems to suggest indigo; but it is too slight to be taken into consideration. Th."

Indigofera hirsuta *L.*—FWTA. ed. 2, 1 : 541 (1958).

Exsicc.: "e Guinea" (**G–DC**—1 sheet, 2: 228. 65, microf. nil; **P–JU**—1 sheet, No. 15386, microf. 1130: III. 7).

I. ferruginea Schum. et Thonn. in Schum. 1827: 370 and 1829: 144; Junghans 1961: 352.

Exsicc.: Ga, Adampi and Augna, *Thonning* 35 (**C**—2 sheets, types, microf. 58: III. 6, 7; 59: I. 1, 2; **S**—1 sheet, type).

"In sandy fields in Ga, Adampi and Augna."

"Differs from *I. hirsuta* in having 2 or 3 pairs of obovate leaflets, much longer and narrower pods, which have a long point and a rust-coloured down. In *I. lateritia* Willd. the leaves are ternate or pinnate, with glutinous hairs, the pods hairy; but Willdenow does not say what colour this down has, and the figure he cites, viz. Jacq. ic. rar. 3 t. 596, more resembles *I. hirsuta* as the leaflets are ovate and the pods beset with white hairs, also the description in Jacq. Collect. 2. p. 359 fits *I. hirsuta* better. S."

"I have nowhere found it in such numbers, that I could institute a regular experiment with it: I collected part of it, removed the stems, and compressed the rest in a very large glass and infused the same with water but it became putrescent without fermenting; neither does the colouring of the leaves and outward aspect indicate any indigo. I doubt whether it has name or utility for the natives, at least I have not been able to discover any of these. Th."

NOTE: This plant is *I. hirsuta* var. *hirsuta*—F.N.H.

Indigofera macrophylla *Schumacher* 1827: 372 and 1829: 146; FWTA. ed. 2, 1: 541 (1958); Junghans 1961: 352.

EXSICC.: Töffri, *Thonning* 86 (**C**—2 sheets, type, microf. 58: III. 3–5); "e Guinea" (**P–JU**—1 sheet, No. 15385, microf. 1130: III. 6; **S**—1 sheet, type).

"In Krepe near the village Töffri."

Indigofera nigricans Vahl ex Pers. 1807: 327 ("Hab. in Guinea. Vahl Hb. Juss."); Poir. 1813: 147; FWTA. ed. 2, 1: 539 (1958); Junghans 1961: 352.

EXSICC.: "e Guinea" (**P–JU**—1 sheet, No. 15371, holotype, microf. 1130: I. 3).

I. elegans Schum. et Thonn. in Schum. 1827: 368 and 1829: 142.

EXSICC.: Ga and Adampi, *Thonning* 61 (**C**—4 sheets, types, microf. 58: II. 3–7; III: 1–2); "e Guinea" (**FI**—1 sheet).

"Grows in Ga and Adampi."

"I communicated this plant to Vahl under the name of *I. nigricans*; meanwhile it does not agree with the character which occurs in Persoon's Synopsis of *I. nigricans*; although it is most likely the same plant which Vahl communicated to Jussieu. Until this doubt is removed I have given it another name. S."

"It is too small to deserve attention as an indigo-plant and the natives do not utilize it. Th."

Indigofera paniculata *Vahl ex Pers.* 1807: 325 ("Vahl in Herb. Juss. Hab. in Guinea"); FWTA. ed. 2, 1: 538 (1958); Junghans 1961: 352.

EXSICC.: "e Guinea" (**P–JU**—1 sheet, No. 15366, holotype, microf. 1129: III. 2).

I. procera Schum. et Thonn. in Schum. 1827: 365 and 1829: 139; FTA. 2: 71 (1871).

EXSICC.: Aquapim, *Thonning* 204 (**C**—3 sheets, types, microf. 57: I. 1–5); "e Guinea" (**C**—1 sheet, microf. 57: I. 6, 7).

"Here and there in valleys near Aquapim."

"It attains a considerable height, but as it constantly loses the older leaves during growth, and the remaining ones are not present in any quantity, it will be unusable for this reason, unless cultivation might make some change; but moisture and drought contribute nothing in this respect, for I found it identical in moist and dry places. Th."

Indigofera pilosa *Poir.* 1813: 151 ("Cette plante croît en Guinée"); FTA. 2: 82 (1871); FWTA. ed. 2, 1: 540 (1958); Junghans 1961: 352.

I. guineensis Schum. et Thonn. in Schum. 1827: 367 and 1829: 141.

Exsicc.: Labode, *Thonning* 307 (**C**—3 sheets, types, microf. 56: III. 3–7 **S**—1 sheet, type); "e Guinea" (**FI**—1 sheet).

"Near Labode in Akkra."

Indigofera pulchra *Willd.* 1802: 1239 ("Habitat in Guinea"); DC. 1825a: 230; Schum. 1827: 369 and 1829: 143; FWTA. ed. 2, 1: 538 (1958); Junghans 1961: 352.

Exsicc.: "Guinea", *Isert* s.n. (**B**—3 sheets, No. 13911, types, microf. 994: II. 9, III. 1, 2; **C**—1 sheet, type, microf. 57: II. 2); Ga and Adampi, *Thonning* 81 (**C**—1 sheet, microf. 57: II. 3, 4; **G–DC**—1 sheet, 2: 230. 86, type, microf. 350. III. 3); "ex Ind. orient" prob. W. African (**C**—1 sheet, microf. 57: II. 1); "e Guinea" (**FI**—1 sheet; **P–JU**—1 sheet, No. 15378, microf. 1130: II. 6).

"In Ga and Adampi."
"The leaves are dried, ground to fine powder and strewn on old leg-injuries, which thereafter are bathed with a decoction of the same. Th."

Indigofera secundiflora *Poir.* 1813: 148 ("Cette plante croît dans la Guinée"); DC. 1825a: 227; FTA. 2: 94 (1871); FWTA. ed. 2, 1: 541 (1958); Junghans 1961: 353. *I. glutinosa* Schumacher 1827: 370 and 1829: 144.

Exsicc.: Atokke, *Thonning* 340 (**C**—3 sheets, types, microf. 56: II. 5–7; III. 1, 2); "e Guinea" (**FI**—1 sheet; **P–JU**—1 sheet, No. 15383, microf. 1130: III. 4; **S**—1 sheet).

"Grows in sand at Atokke in Augna."

Indigofera spicata *Forsk.*—FWTA. ed. 2, 1: 542 (1958). *I. anceps* Poir. 1813: 147 ("Cette plante croît dans la Guinée"); Junghans 1961: 351.

I. hendecaphylla Jacq. 1788: 358; 1786–93b: pl. 57a; Willd. 1802: 1233; Schum. 1827: 374 and 1829: 148, as to Thonning specimens; FTA. 2: 96 (1871).

Exsicc.: Tessin and Quitta, *Thonning* 42 (**C**—7 sheets, types, 59: III. 2, 3, 5–7; 60: I. 1–6; **G–DC**—1 sheet, 2: 228. 72, type, microf. 350: II. 5); "e Guinea" (**C**—1 sheet, microf. 59: III. 1; **FI**—1 sheet; **P–JU**—1 sheet, No. 15376, type, microf. 1130: II. 4).

"Rare; at Tessin and Quitta."
"Jacquin's description and illustration do not fit well. S."
"Its outward appearance betrays no indigo; the leaves are of an ordinary dark-green colour. Th."

I. thonningii sensu Junghans 1961: 353, pro parte non Vahl ex Poir.

Exsicc.: "Guinea", *Thonning* "45" (**C**—1 sheet, microf. 59: III. 4).

Indigofera subulata *Vahl ex Poir.* 1813: 150 ("Cette plante croît dans la Guinée"); FTA. 2: 87 (1871); Meikle in Kew Bull. 5: 351 (1951); FWTA. ed. 2, 1: 541 (1958); Junghans 1961: 353.

I. thonningii Schum. et Thonn. in Schum. 1827: 366 and 1829: 140.

Exsicc.: Labodei to Ursue, *Thonning* 45 (**C**—4 sheets, type, microf. 57: II. 5–7; III. 1–4; **G–DC**—1 sheet, 2: 223. 14, microf. 347: III. 7); "e Guinea" (**FI**—1 sheet; **P–JU**—1 sheet, No. 15384, microf. 1130: III. 5; **S**—1 sheet, type).

"In a single spot between Labodei and Ursue in Ga."

Note (i). Our plant is *I. subulata* var. *subulata*—F.N.H.

Note (ii). A sheet at Copenhagen numbered "45" and apparently included by Schumacher with the above is referable to *I. spicata* (q.v.).—F.N.H.

Indigofera tetrasperma *Vahl ex Pers.* 1807: 325 ("Vahl Hab. in Guinea. (Hb. Juss.)"); Poir. 1813: 150; DC. 1825a: 222; Schum. 1827: 365 and 1829: 139; FWTA. ed. 2, 1: 539 (1958); Junghans 1961: 353. *I. scoparia* Vahl ex DC. 1825a: 222; Junghans 1961: 353.

Exsicc.: "Guinea", 1786, *Isert* s.n. (**C**—1 sheet, microf. 58: I. 1); Ada, 1784, *Isert* s.n. (**C**—1 sheet, microf. 58: II. 1, 2); Ga, *Thonning* 41 (**C**—4 sheets, types, microf. 57: III. 7; 58: I. 2–7); "e Guinea" (**FI**—1 sheet; **G–DC**—1 sheet, 2: 222. 2, microf. 347: II. 4; **P–JU**—1 sheet, No. 15365A, holotype, microf. 1129: II. 7; **S**—1 sheet).

"Not common; in cultivated places, e.g. in Ga."
"It is so rare that I have not had an opportunity of testing it as regards its use; it does not deserve much attention in this respect as it is so small and has only a few leaves. Th."

Indigofera tinctoria *L.*—FTA. 2: 99 (1871); FWTA. ed. 2, 1: 541 (1958).

I. ornithopodioides Schumacher 1827: 372 and 1829: 146; Junghans 1961: 352.

Exsicc.: Ga, Adampi and Augna, *Thonning* 20 (**C**—1 sheet, type, microf. 60: I. 7; II. 1).

"In Ga, Adampi, Augna near the villages."
"Presumably different from *I. tinctoria* and *I. anil*; in *I. tinctoria* the leaflets are of the same shape as well, but the pods are indistinctly four-angled and not knotted; in *I. anil* the leaflets are more oblong-ovate; the pods sickle-shaped, indistinctly 4-angled and only half as long. S."
"This species of Indigofera is the commonest of the known species in Guinea; it grows by every native dwelling which is at the seaside, whereas I have not found it in fields at some distance from the dwellings, not at all at Rio Volta, when I except some few plants which came from seeds, which were ejected onto the soil. At Ningo and Quitta I found it in the greatest quantity and at the same time in the best growth; it seems to be partial to a sandy, not all too hard, soil; it can endure much drought for at the driest time of year, when most plants fade, this plant is still in good growth; it flowers nearly the whole year through but mostly after continuous rain, it is not much plagued with insects and is not eaten by domestic creatures. Although this species of Indigofera is the commonest and largest, it is nevertheless nowhere present in sufficient quantity that it could sufficiently provide an indigo-factory without special cultivation. In order to find out how it will

thrive in a rich and moist soil, which is just the opposite of that in which it grows, I had some seeds sown in the settlements at Dudna. I have not been able to discover if the natives know of any medicinal or economic use for this plant, I have not at all been able to find out its name, as it nevertheless is a plant which grows at their dwellings; and the only use which is made of it at times, is as a besom. Th.''

Lablab purpureus (*L.*) *Sweet*—FTEA. Legum.—Phaseol. 696 (1971); Hepper in Kew Bull. 26: 565 (1972).

Dolichos nervosus Schum. et Thonn. in Schum. 1827: 342 and 1829: 116 (no MS No.); Junghans 1961: 349.

Exsicc.: Ga, *Thonning* s.n. (**C**—1 sheet, microf. 37: I. 5).

"In Ga.''

Lonchocarpus cyanescens (*Schum. et Thonn.*) *Benth.*—FTA. 2: 243 (1871); FWTA. ed. 2, 1: 523 (1958).

Robinia cyanescens Schum. et Thonn. in Schum. 1827: 351 and 1829: 125; Junghans 1961: 354.

Exsicc.: "Guinea'', *Thonning* 77 (**C**—4 sheets, types, microf. 92: II. 5–7; III. 1–3); "e Guinea'' (**P–JU**—1 sheet, No. 15224, microf. 1120: I. 6).

Vernacular name: 'Akassi'.

"Grows here and there in fields.''
"The root is used as in the foregoing [*Robinia multiflora*]. The leaves are crushed and applied to old leg-injuries for cleansing the wound. Th.''

Lonchocarpus sericeus (*Poir.*) *Kunth*—FTA. 2: 241 (1871); FWTA. ed. 2, 1: 522 (1958).

Robinia argentiflora Schum. et Thonn. in Schum. 1827: 352 and 1829: 126; Junghans 1961: 354.

Exsicc.: Ada, 1784, *Isert* s.n. (**C**—1 sheet, microf. 92: II. 4, fruits are of *Millettia* sp.); Ada, *Thonning* 166 (**C**—3 sheets, types, microf. 92: II. 1–3); "e Guinea'' (**P–JU**—1 sheet, No. 15225, microf. 1120: I. 7).

Vernacular name: 'Lablaku'.

"At Ada''.
"It approaches *R. mollis* Rohr from America; the difference is that in *R. mollis* the branches are streakedly beset with a rust-coloured felt; in *R. argentiflora* on the other hand almost smooth beset with pale points; in the former the racemes are longer and more sloping and also the leaf-stalks more densely felted, in the latter the racemes are erect, stiff and also the leaf-stalks less felted; in the former the lowest leaflets are ovate pointed, in the latter nearly round. Yet it is quite possible that these two plants are only subspecies. S.''

Macrotyloma biflorum (*Schum. et Thonn.*) *Hepper* in Kew Bull. 26: 565

(1972). *Dolichos chrysanthus* A. Chev.—FWTA. ed. 2, 1: 571 (1958). *Macroty-loma chrysanthum* (A. Chev.) Verdc. in Kew Bull. 24: 402 (1970).

Glycine biflora Schum. et Thonn. in Schum. 1827: 345 and 1829: 119 (no MS No.); Junghans 1961: 350.

Exsicc.: "Guinea", *Thonning* s.n. (**C**—4 sheets, types, microf. 45: I. 5–7; II. 1–4); "e Guinea" (**C**—1 sheet, microf. 45: I. 4; **FI**—1 sheet; **P–JU**—1 sheet, No. 15179, microf. 1116: II. 3).

"Here and there."

var. "*β*" "Foliis longioribus angustioribus, acutioribus."

NOTE: This variety was given the above description but no name; neither was a specimen cited, so the type is presumably one of those at Copenhagen cited above.—F.N.H.

Millettia irvinei *Hutch. et Dalz.*—FWTA. ed. 2, 1: 526 (1958).

Robinia multiflora Schum. et Thonn. in Schum. 1827: 350 and 1829: 124; Junghans 1961: 354, non *M. multiflora* Coll. et Hemsl.

Exsicc.: "Guinea", 1786, *Isert* s.n. (**C**—1 sheet, microf. 92: III. 6); "Guinea", *Thonning* 108 (**C**—6 sheets, types, microf. 92: III. 4, 5, 7; 93: I. 1–7; II. 1); "e Guinea' (**P–JU**—1 sheet No. 15226, microf. 1120: II. 1).

VERNACULAR NAME: 'Ahaemeté'.

"Not very common. Flowers in June."
"The root is beaten out to a soft sponge which the natives use for washing with. Th."

NOTE: *M. irvinei* was a *nom.* nov. since the epithet 'multiflora' was already used in Millettia.—F.N.H.

Millettia thonningii (*Schumacher*) *Baker*—FTA. 2: 128 (1871); FWTA. ed. 2, 1: 527 (1958).

Robinia thonningii Schumacher. 1827: 349 and 1829: 123; Junghans 1961: 354.

Exsicc.: Ada, 1784, *Isert* s.n. (**C**—1 sheet, microf. 94: I. 1, 2); Volta R., *Thonning* 15 (**C**—6 sheets, types, microf. 93: II. 2–7; III. 1, 4–7); "e Guinea" (**C**—1 sheet, microf. 93: III. 2, 3).

VERNACULAR NAME: 'Tah-tjo'.

"Grows here and there in the field but mainly near the Volta river."
"The tree is almost the same size as a beech. The wood is pretty hard and whitish yellow, but not very fine. The bark beaten soft is used for applying to old leg-injuries for cleansing the wound. Th."

Mucuna sloanei *Fawc. et Rendle*—FWTA. ed. 2, 1: 561 (1958). *M. urens* DC.—FTA. 2: 185 (1871).

Stizolobium urens (L.) Pers.—Schum. 1827: 343 and 1829: 117; Junghans 1961: 354.

Exsicc.: Aquapim, *Thonning* 146 (**C**—2 sheets, microf. 105: II. 1–3).

Vernacular name: 'Taetjoe-pang".

"In Aquapim, Volta-Krepé."
"The natives crush the stem and leaves and colour leather black with the expressed sap, merely in smearing it over and letting it dry in the air. Th."

Ophrestia hedysaroides (*Willd.*) *Verdc.* in Kew Bull. 24: 259 (1970); FTEA. Legum.—Papil. 526 (1971).

Glycine hedysaroides Willd. 1802: 1060 ("Habitat in Guinea"); Schum. 1827: 345 and 1829: 119; FWTA. ed. 2, 1: 564 (1958); Junghans 1961: 350.

Exsicc.: "Guinea", *Isert* s.n. (**B**—1 sheet, No. 13437, type, microf. 958: III. 9; **C**—1 sheet, type, microf. 45: III. 4); "Guinea", *Thonning* 174 (**C**—4 sheets, microf. 45: II. 5–7; III. 1–3, 5); "e Guinea"? (**C**—1 sheet, microf. 45: III. 6).

"Grows here and there. Flowers in May and June."

Papilionaceae indet. in Junghans 1961: 355.

Exsicc.: "Guinea: sylvaticis montosis," July 1786, *Isert* s.n. (**C**—1 sheet, microf. 64: II. 4).

Ormocarpum sennoides (*Willd.*) *DC.* subsp. **hispidum** (*Willd.*) *Brenan et J. Léonard*—FWTA. ed. 2, 1: 576 (1958). *O. sennoides* (Willd.) DC.—FTA. 2: 143 (1871). *Robinia guineensis* Willd. 1809: 769.

Cytisus hispidus Willd. 1802: 1121 ("Habitat in Guinea"); Junghans 1961: 348.

Exsicc.: "Guinea", *Isert* s.n. (**B**—1 sheet, No. 13645, type, microf. 974: II. 6; **C**—1 sheet, type, microf. 91: II. 6, 7).

Rathkea glabra Schum. et Thonn. in Schum. 1827: 355 and 1829: 129.

Exsicc.: Ga, *Thonning* 47 (**C**—6 sheets, type, microf. 91: I. 1–7, II. 1–3); "e Guinea" (**C**—1 sheet, microf. 91: II. 4–5; **FI**—1 sheet).

"Not common, in Ga; flowers in June."
"*R. squamata* Vahl, Symb. 3. p. 88 t. 69; Willd., Sp. pl. p. 1133, *Hedysarum sennoides* Willd., Sp. pl. 3. p. 1207, (*H. fruticosum* Röttler) and several more belong to this genus. S."
"I have named this genus after *Rathke*, Professor of Natural History in the University of Christiania. S."

Phaseolus adenanthus *G. F. W. Mey.*—FWTA. ed. 2, 1: 365 (1958); Hepper in Kew Bull. 26: 566 (1972).

Dolichos oleraceus Schum. et Thonn. in Schum. 1827: 340 and 1829: 114 (MS No. 145); Junghans 1961: 349.

Exsicc.: Christiansborg, 1784, *Isert* s.n. (**C**—1 sheet, type, microf. 37: II. 4–5).

Vernacular name: "Jo"; "Quitto-bönner" (Danish); "Calevancus" (English).

"Cultivated."

"The Augna natives frequently cultivate this plant; each plant spreads out about 16 square ells and yields a rich harvest. The pods must be plucked from time to time when they are ripe, and every half year they must be sown anew. The beans are of the size of the common bean (*Phaseolus vulgaris*) and fairly good-tasting; they are sold as cabin-provision for the ships. The natives use the leaves as cabbage and a careful use of the plant in this way will give much more fruit. Th."

Note: This has previously been considered to be a synonym of *Vigna unguiculata* (L.) Walp. and there may be some confusion in Thonning's description which partly applies to it and partly to *P. adenanthus*. Isert's specimen is certainly of the latter.—F.N.H.

Phaseolus lunatus *L.*—FWTA. ed. 2, 1 : 565 (1958).

Papilionaceae indet. in Junghans 1961 : 355.

Exsicc.: Whydah, 1785, *Isert* s.n. (**C**—1 sheet, microf. 79: II. 7; III. 1).

Phaseolus vulgaris *L.*—Schum. 1827: 338 and 1829: 112 (no MS No.); Junghans 1961 : 353.

"Cultivated".

Note: The Isert specimen referred to by Junghans l.c. is probably the one now placed under *P. lunatus*.—F.N.H.

Pseudovigna argentea *(Willd.) Verdc.* in Kew. Bull. 24: 352, fig. 4 (1970); FTEA. Legum.—Papil. 598 (1971).

Dolichos argenteus Willd. 1802: 1047 ("Habitat in Guinea"); FWTA. ed. 2, 1 : 571 (1958); Junghans 1961 : 349.

Exsicc.: Whydah, *Isert* (**B**—1 sheet, No. 13411, type, microf. 957: II. 4; no type material at **C**).

Glycine dentata Vahl ex Schum. 1827: 348 and 1829: 122.

"At Fida according to Isert." (See above).

Pterocarpus santalinoides *L' Hérit. ex DC.*—FWTA. ed. 2, 1 : 517 (1958).

P. esculentus Schum. et Thonn. in Schum. 1827: 330 and 1829: 104 (MS No. 344); Junghans 1961 : 354.

Exsicc.: Ada, 1784, *Isert* s.n. (**C**—1 sheet, microf. 90: III. 3); "India occid." (see note below) (**C**—1 sheet, microf. 90: III. 4, 5).

Vernacular name: 'Gaegaenae'.

"Grows plentifully on the banks of the river Volta."

"The unripe fruit contains a white pulp which encloses the seeds; this is roasted and eaten by the natives."

D

NOTE: No specimen is labelled as Thonning's, although the sheet written up as "India occid." may also be his. It cannot however be No. 344 (according to the MS) as this number is used for *Setaria pallide-fusca*—F.N.H.

Rhynchosia minima (*L.*) *DC.*—FTA. 2: 219 (1871); FWTA. ed. 2, 1: 555 (1958).

Glycine rhombea Schum. et Thonn. in Schum. 1827: 346 and 1829: 120; Junghans 1961: 350.

EXSICC.: "Guinea", *Thonning* 194 (**C**—3 sheets, types, microf. 46: I. 7; II. 1–4); "e Guinea" (**FI**—1 sheet; **P–JU**—1 sheet, No. 15215, microf. 1119: II. 1).

"In dry fields along the ground and twining around species of grass; flowers in July."

"It is rare to find a species of Glycine without (sic) the resinous dots on the under-surface of the leaves. S."

Rhynchosia pycnostachya (*DC.*) *Meikle* in Kew Bull. 9: 274 (1954); FWTA. ed. 2, 1: 554 (1958); Hepper in Kew Bull. 26: 566 (1972).

Glycine macrophylla Thonning in Schum. 1827: 348 and 1829: 122 (no MS No.); Junghans 1961: 350.

EXSICC.: "Guinea", *Isert* s.n. (**C**—1 sheet, microf. 46: I. 1, 2); "Guinea", *Thonning* s.n. (**C**—1 sheet, type, microf. 45: III. 7); "e Guinea" (**C**—2 sheets, microf. 46: I. 3–6). (See note below).

"Wonder if *G. caribaea*? The leaf-shape answers well to this plant. S."

NOTE: One of these sheets (microf. 46: I. 5, 6) is labelled "e Guinea ded. Dr. Banks" which is one of two references to Banks in this collection (vide *Melastomastrum capitatum*). Since there are Smeathman specimens of each species at the British Museum I suggest they are duplicates of his collections, as is certainly the case with *Afzelia parviflora* (q.v.).—F.N.H.

Rhynchosia sublobata (*Schum. et Thonn.*) *Meikle* in Kew Bull. 6: 176 (1951); FWTA. ed. 2, 1: 555 (1958). *R. caribaea* (Jacq.) DC.—FTA. 2: 220 (1871).

Glycine sublobata Schum. et Thonn. in Schum. 1827: 347 and 1829: 121; Junghans 1961: 350.

EXSICC.: Christiansborg, 1784, *Isert* s.n. (**C**—1 sheet, microf. 46: III. 5, 6); "Guinea", *Thonning* 247 (**C**—3 sheets, types, microf. 46: III. 7; 47: I. 1–4, 6); "e Guinea" (**P–JU**—1 sheet, No. 15180, microf. 1116: II. 4); "ex Ind. orient" (see note below) (**C**—1 sheet, microf. 47: I. 5).

VERNACULAR NAME: 'Nanni-jaa'.

"Here and there in open, high and poor fields."

NOTE: R. D. Meikle wrote on the "ex Ind. orient" sheet in 1948 that the locality is an error: "this is almost certainly a portion of Thonning's Guinea material".—F.N.H.

Sesbania pachycarpa *DC. emend. Guill. et Perr.*—Gillett in Kew Bull. 17: 126 (1963). *S. bispinosa* sensu FWTA. ed. 2, 1 : 532, non (Jacq.) W. F. Wight.

Emerus aculeata (Willd.) Hornem. 1815: 696; 1819: 23; Schum. 1827: 353 and 1829: 127; Junghans 1961 : 349.

Exsicc.: Ga, *Thonning* 304 (**C**—4 sheets, microf. 39: II. 5-7; III. 1-5); "e Guinea" (**P-JU**—1 sheet, No. 15512, microf. 1141 : I. 2; **FI**—1 sheet).

"In Ga in the fields."

Sesbania sericea *(Willd.) Link*—Gillett in Kew Bull. 17: 133 (1963). *S. pubescens* DC. 1825a: 265; FWTA. ed. 2, 1 : 532 (1958); Junghans 1961 : 354.

Emerus pubescens (DC.) Schum. et Thonn. in Schum. 1827: 354 and 1829: 128.

Exsicc.: Ga, Adampi, Augna, *Thonning* 43 (**C**—3 sheets, types, microf. 39: III. 6, 7; 40: I. 1-3); "e Guinea" (**P-JU**—1 sheet, No. 15513, microf. 1141 : I. 3).

"Not rare, in Ga, Adampi, Augna."
"Closely related to the foregoing [*S. pachycarpa*] but sufficiently different in the blue-green felted leaves and the rather angled downy-haired stem. Differs from *Coronilla picta* Willd. in the prickly spike- and leaf-stalks. S."

Sophora occidentalis *L.*—FWTA. ed. 2, 1 : 509 (1958). *S. tomentosa* L.—FTA. 2 : 254 (1871).

S. nitens Schum. et Thonn. in Schum. 1827: 201 and 1828: 221; Junghans 1961 : 354.

Exsicc.: R. Volta mouth and Poisi, *Thonning* 113 (**C**—1 sheet, type, microf. 103: III. 4, 5).

"Here and there amongst the shrubs at the mouth of the river Volta; similarly at Poisi. It grows in loose sandy ground, and flowers in October and May."
"Much resembles *S. orientalis* L. but in this the fine felt which covers all parts has a less silvery sheen, all leaflets are alternate, rather distant and longer stalked; lomenta have a much longer and narrower point. S."

Stylosanthes erecta *P. Beauv.*—FTA. 2 : 156 (1871); FWTA. ed. 2, 1 : 575 (1958).

S. guineensis Schum. et Thonn. in Schum. 1827: 357 and 1829: 131; Junghans 1961 : 354.

Exsicc.: Whydah, 1785, *Isert* s.n. (**C**—1 sheet, microf. 105: III. 5, 6); "Guinea" coast, *Thonning* 231 (**C**—3 sheets, type, microf. 105: III. 4; 106: I. 1, 2; one of the three sheets not photographed); "e Guinea" (**P-JU**—1 sheet No. 15636, microf. 1149: III. 3).

"Very common in the fertile season of the year in poor spots near the sea."
"*S. erecta* Beauv. Fl. d'Oware and de Benin Tab. IXXVII only differs in having smooth bracts; so that perhaps it is the same plant."

Tephrosia bracteolata *Guill. et Perr.*—FWTA. ed. 2, 1: 530 (1958).

Papilionaceae indet. in Junghans 1961: 355.

EXSICC.: Ada, 1784, *Isert* s.n. (**C**—1 sheet, microf. 43: III. 2).

Tephrosia elegans *Schumacher* 1827: 376 and 1829: 150 (no MS No.); FWTA. ed. 2, 1: 529 (1958); Junghans 1961: 354.

EXSICC.: Aquapim, *Thonning* s.n. (**C**—2 sheets, types, microf. 106: I. 6, 7); without data (**C**—1 sheet, microf. 106: II, 1).

"In Aquapim."

Tephrosia linearis (*Willd.*) *Pers.* 1807: 330; Schum. 1827: 378 and 1829: 152 (no MS No.); FWTA ed. 2, 1: 531 (1958); Junghans 1961: 355.

Galega linearis Willd. 1802: 1248 ("Habitat in Guinea.")

EXSICC.: Ada, 1784, *Isert* s.n. (**C**—1 sheet, type, microf. 43: II. 7; III. 1; **B**—1 sheet, No. 13941, type, microf. 996: II. 3); Ga, Adampi, Augna, *Thonning* s.n. (**C**—2 sheets, microf. 43: II. 3, 5, 6); "e Guinea" (**C**—1 sheet, microf. 43: II. 4).

"Frequent in Ga, Adampi, Augna."

Tephrosia pumila (*Lam.*) *Pers.*—FTEA. Legum.—Papil. 184 (1971); Hepper in Kew Bull. 26: 566 (1972). *T. uniflora* sensu FWTA. ed. 2, 1: 529 (1958).

T. hirsuta Schum. et Thonn. in Schum. 1827: 377 and 1829: 151 (no MS. No.); Junghans 1961: 354.

EXSICC.: Ga and Adampi, *Thonning* s.n. (**C**—2 sheets, types, microf. 106: II. 2–4); "e Guinea" (**P–JU**—1 sheet, No. 15356, microf. 1129: I. 1).

"In Ga and Adampi."

Tephrosia pupurea (*L.*) *Pers.* subsp. **leptostachya** (*DC.*) *Brummitt* var. **leptostachya**—Bol. Soc. Brot., sér. 2, 41: 245 (1968); FTEA. Legum.—Papil. 186 (1971).

T. lineata Schum. et Thonn. in Schum. 1827: 376 and 1829: 150 (no MS. No.); Junghans 1961: 355.

EXSICC.: "Guinea", *Thonning* s.n. (**C**—2 sheets, types, microf. 106: II. 5, 7; III. 1); "e Guinea" (**P–JU**—1 sheet, No. 15357, microf. 1129: I. 2).

"Related to *T. purpurea* but the latter is quite smooth, has much larger leaflets, much longer many-flowered racemes, larger closer-placed flowers, very fine flower-stalks of 2 lines in length; where as *T. lineata* is beset with soft hairs, the leaflets are smaller, the racemes shorter, the flowers smaller, more distant, the flower-stalks not a line long, thicker. S."

Uraria picta (*Jacq.*) *DC.*—FTA. 2: 169 (1871); FWTA. ed. 2, 1: 587 (1958).

Hedysarum pictum Jacq. 1788: 262; 1786–1793b: pl. 567 (". . . Guineae arenosis . . . Iserto detecta . . .''); Willd. 1802: 1204; Schum. 1827: 364 and 1829: 138; Junghans 1961: 351.

Exsicc.: "Guinea", *Isert*, s.n. (**B**—2 sheets, No. 13834, types, microf. 987: III. 9; 988: I. 1; **C**—1 sheet, type (see note), microf. 51: III. 5, 6); "Guinea", *Thonning* 129 (**C**—1 sheet, microf. 51: III. 4); "e Guinea" (**FI**—1 sheet; **P–JU**—1 sheet, No. 15579, microf. 1146: I. 3).

"Here and there not rare, especially in open fields."

Note: The holotype of *H. pictum* is not in Vienna, although the plant was raised from seed obtained from Isert. The latter's specimens elsewhere could be considered isotypes.—F.N.H.

Vigna ambacensis *Baker*—FWTA. ed. 2, 2: 568 (1958).

Papilionaceae indet. in Junghans 1961: 355.

Exsicc.: Ada, 1784, *Isert*, s.n. (**C**—1 sheet, microf. 109: I. 6–7).

Vigna luteola (*Jacq.*) *Benth.*—FWTA. ed. 2, 1: 569 (1958).

Papilionaceae indet. in Junghans 1961: 355.

Exsicc.: Ada, 1784, *Isert* s.n. (**C**—1 sheet, microf. 37: III. 2, 3).

Vigna vexillata (*L.*) *A. Rich.*—FTA. 2: 199 (1871); FWTA. ed. 2, 1: 567 (1958). *Dolichos angustifolius* Vahl ex Guill. et Perr.—1832: 220.

Plectrotropis angustifolia Schum. et Thonn. in Schum. 1827: 338 and 1829: 112; Junghans 1961: 353.

Exsicc.: Quitta, *Thonning* 7 (**C**—2 sheets, types, microf. 85: II. 2–4).

"Grows in a few places, mainly in neighbourhood of Quitta amongst the tall grass, which it becomes twined around."

Plectrotropis hirsuta Schum. et Thonn. in Schum. 1827: 339 and 1829: 113.

Exsicc.: Aquapim, *Thonning* 213 (**C**—1 sheet, type, microf. 85: II. 5).

"Grows like the foregoing [*P. angustifolia*], but nearer Aquapim in the valleys."
"The generic name I formed from Greek πλῆκτρον (plectron) *calcar* and τρόπις (tropis) *carina*. S."

Note: As *D. angustifolius* was not validly published by Vahl, the author citation strictly should be (Schum. et Thonn.) Guill. et Perr.—F.N.H.

Voandzeia subterranea (*L.*) *Thouars*—FWTA. ed. 2, 1: 572 (1958).

Glycine subterranea L.—Schum 1827: 344 and 1829: 118 (no MS No.); Junghans 1961: 350.

Exsicc.: "Guinea", *Isert* s.n. (**C**—1 sheet, microf. nil); "Guinea", *Thonning* s.n. (**C**—1 sheet, microf. nil); "e Guinea"? (**C**—1 sheet, microf. nil).

Vernacular name: 'Aquing' (Thonning); 'Jubbejubbe' (Isert).

"Here and there."

Zornia glochidiata *Reichb. ex DC.*—FWTA. ed. 2, 1 : 575 (1958).

Z. diphylla sensu Schum. 1827: 358 and 1829: 132 (no MS No.); Junghans 1961 : 355, non (L.) Pers.

EXSICC.: Ada, 1784, *Isert* s.n. (**C**—1 sheet, microf. 112 : II. 1, 2); "Guinea", *Thonning* s.n. (**C**—2 sheets, microf. 112 : I. 5–7).

"Wonder if all the species of *Zornia* cited by Persoon are correctly separated from *Hedysarum*? *Zornia pulchella* P(ers). (*Hedysarum pulchellum* L.) has not stiff-haired hardly downy lomenta. S."

PASSIFLORACEAE

Adenia lobata (*Jacq.*) *Engl.*—FWTA. ed. 2, 1 : 203 (1954). *Modecca lobata* Jacq.—FTA. 2 : 516 (1871).

Modekka diversifolia Schum. et Thonn. in Schum. 1827: 435 and 1829: 209; Junghans 1962 : 82.

EXSICC.: Accra, 1786, *Isert* s.n. (**C**—1 sheet, microf. 77: III. 5); "Guinea", *Thonning* 250 (no type material at **C**; **S**—2 sheets   , types).

VERNACULAR NAME: 'Koo-pang'.

"Here and there among the bushes."
"*Modecca* Lamarck Dict. Bot. 4, p. 208 and Vahl Naturhistorie-Selskabets publications 6 part p. 103, is the mentioned genus here. The figure in Rheede Hortus malabaricus 8. t. 20–23 belongs to the same genus, but to another species. In appearance and shape of flower this genus approaches *Passiflora*. S."

Smeathmannia pubescens *Soland. ex R. Br.*—FTA. 2: 506 (1871); FWTA. ed. 2, 1 : 200 (1954).

Buelowia illustris Schum. et Thonn. in Schum. 1827: 247 and 1829: 21; Junghans 1962 : 82.

EXSICC.: *Thonning* 85 (no type material traced).

"The genus is named after the late Privy Conference Councillor Johan von Bülow, who in Denmark has so greatly supported the sciences and amongst these Botany in particular. S."

PEDALIACEAE

Sesamum alatum *Thonning* in Schum. 1827: 284 and 1829: 58; FWTA. ed. 2, 2 : 389 (1963); Junghans 1962 : 82.

EXSICC.: "Guinea", *Thonning* 277 (**C**—1 sheet, type, microf. 98: I. 7; II. 1).

VERNACULAR NAME: 'Otru'.

"Here and there."
"Made use of as the foregoing" [i.e. *S. radiatum* below].

Sesamum radiatum *Schum. et Thonn.* in Schum. 1827: 282 and 1829: 56; FWTA. ed. 2, 2: 391 (1963); Junghans 1962: 82.

Exsicc.: "Guinea", *Thonning* 276 (C—1 sheet, type, microf. 98: II. 2, 3); "e Guinea" (P–JU—1 sheet, No. 4944A, microf. 341: I. 5).

Vernacular name: 'Otru'.

"In cultivated and fertile places."

"The leaves are used as cabbage either by themselves or mixed with others. Th."

"Related to *S. orientale*, but in the latter the whole plant is less hairy, the leaves entire-margined, the leaf-stalks longer (nearly 1 inch long), the flowers white, the flower-stalks shorter, finer, cylindric subtended by 2 bracts; and the capsule is provided with a sharp point. S."

PERIPLOCACEAE see ASCLEPIADACEAE

PHYTOLACCACEAE

Hilleria latifolia *(Lam.)* *H. Walt.*—FWTA. ed. 2, 1: 143 (1954); FTEA. Phytolacc. 6 (1971). *Mohlana nemoralis* Mart.—FTA. 6, 1: 95 (1909).

Rivina apetala Schum. et Thonn. in Schum. 1827: 84 and 1828: 104; Junghans 1962: 83.

Exsicc.: Aquapim, 1786, *Isert* s.n. (C—1 sheet, microf. 92: I. 6); Aquapim, *Thonning* 259 (C—2 sheets, types, microf. 92: I. 5, 7; P–JU—1 sheet, type, No. 4390, microf. 297: III. 4).

"Frequent in Aquapim."

"*R. laevis* differs from our plant in having ovate nearly cordate leaves, and 4-leaved corollas. *R. secunda* Flor. Per. has ovate pointed, indistinctly denticulate leaves. S."

PIPERACEAE

Piper guineense *Thonning* in Schum. 1827: 19 and 1828: 39; FWTA. ed. 2, 1: 84 (1954); Junghans 1962: 83.

Exsicc.: Akvapim etc., *Thonning* 282 (C—4 sheets, types, microf. 85: I. 1–6).

Vernacular name: 'Dojvié'; 'Asiante-Peber' (Danish).

"Grows in Akvapim, Asiante, Akvambu and Krepe in moist woods; it twines around tree stems to a height of nearly 24 ft., and unites very closely with the same, throwing out small roots, which become attached to the bark, or to the moss clothing it. I have not myself found the plant in its habitat, but according to the natives' statement, the leaves are tightly applied to the surface of the tree; some state that the catkins are single, others that they are aggregated together."

"The fruit is a bad substitute for black pepper, as it has an unpleasant bitterness, which is stronger and more distasteful the fresher it is. Th."

PLUMBAGINACEAE

Plumbago zeylanica *L.*—FWTA. ed. 2, 2: 306 (1963).

P. auriculata Lam.—Schum. 1827: 88 and 1828: 108; Junghans 1962: 83.

Exsicc.: "Guinea", *Thonning* 7 (**C**—5 sheets, microf. 85: II. 6, 7; III. 1–7).

"Fairly common in thickets."
"The natives do not know the medicinal use of this plant as a vesicatorium."

POLYGALACEAE

Polygala arenaria *Willd.* 1802: 880 ("Habitat in Guinea"); Schum. 1827: 328 and 1829: 102; FWTA. ed. 2, 1: 112 (1954); Junghans 1962: 83.

Exsicc.: Whydah, 1785, *Isert* s.n. (**B**—1 sheet, No. 12977, type, microf. 930: III. 2; **C**—1 sheet, type, microf. 87: III. 5); "Guinea", *Thonning* 70 (**C**—6 sheets, microf. 87: III. 4; 88: I. 1–7; II. 1–3); "e Guinea" (**C**—2 sheets, microf. 87. III. 3, 6, 7; **P–JU**—1 sheet, No. 12850, microf. 949: III. 3).

"Here and there in the higher fields near the coast. Flowers in June."

Polygala guineensis *Willd.* 1802: 882 ("Habitat in Guinea"); Schum. 1827: 329 and 1829: 103; FWTA. ed. 2, 1: 112 (1954); Junghans 1962: 83.

Exsicc.: "Guinea", 1786, *Isert* s.n. (**B**—1 sheet, No. 12980, type, microf. 930: III. 6; **C**—1 sheet, type, microf. 88: II. 4); Christiansborg, *Thonning* 71 (**C**—4 sheets, microf. 88: II. 5–7; III. 1–5).

"In neighbourhood of Christiansborg; flowers in May."

NOTE: One of the Thonning sheets bears the number 66 and none of the others are numbered, but I feel sure that they should all be regarded as No. 71.—F.N.H.

PORTULACACEAE

Portulaca foliosa *Ker-Gawl.*—FTA. 1: 148 (1868); FWTA. ed. 2, 1: 137 (1954).

P. prolifera Schum. et Thonn. in Schum. 1827: 239 and 1829: 13; Junghans 1962: 83.

Exsicc.: "Guinea", *Thonning* 102 (no type material traced).

ROSACEAE

Chrysobalanus orbicularis *Schumacher* 1827: 232 and 1829: 6 (no MS); FWTA. ed. 2, 1: 426 (1958); Junghans 1962: 83. *C. icaco* sensu FTA. 2: 365 (1871), non L.

Exsicc.: "Guinea", *Thonning* 54 (**C**—2 sheets, types, microf. 19: III. 7; 20: I. 1–4); "e Guinea"? (**C**—4 sheets, microf. 19: III. 2–6; 20: I. 5).

Note: Junghans attributes one of these sheets to Isert—F.N.H.

"It resembles *C. icaco* which yet differs from it in that the leaves are wedge-shaped almost obovate, racemi paniculati, the flower-stalks longer and out-spread, also the branches are beset with larger light-brown "Hysterier." S."

Note: W. C. Worsdell comments: "These Hysterii are conspicuous lenticels in form of uteri (hysterii)."—F.N.H.

RUBIACEAE

Benzonia corymbosa *Schumacher* 1827: 114 and 1828: 134 (no MS.); FTA. 3: 247 (1871); Junghans 1962: 84.

Exsicc.: *Thonning* s.n. (no type material at **C**).

"I have called this genus after Peder Eggert Benzon, chemist in St. Croix, who has collected many plants on the Danish West Indian islands and written a paper on the West Indian Salop. S."

Note: It is extraordinary that this plant has not been identified. Unfortunately no material has been traced and neither Dr. B. Verdcourt nor I (in FWTA. ed. 2, 2: 110 (1963)) have been able to suggest its identity—F.N.H.

Borreria filifolia *(Schum. et Thonn.) K. Schum.*—FWTA. ed. 2, 2: 221 (1963).

Octodon filifolium Schum. et Thonn. in Schum. 1827: 74 and 1828: 94; FTA. 3: 241 (1877); Junghans 1962: 86.

Exsicc.: Ningo, *Thonning* 241 (**C**—3 sheets, types, microf. 71: I. 4–7; II. 1); "Guinea" (**FI**—1 sheet).

"Occurs at Ningo, although rarely."

Borreria scabra *(Schum. et Thonn.) K. Schum.*—FWTA. ed. 2, 2: 220 (1963).
Spermacoce ruelliae DC.—FTA. 3: 238 (1877).

Diodia scabra Schum. et Thonn. in Schum. 1827: 76 and 1828: 96; Junghans 1962: 84.

Exsicc.: "Guinea" coast, *Thonning* 191 (**C**—1 sheet, type, microf. 36: I. 6); "e Guinea" (**C**—1 sheet, microf. 36: I. 7; II. 1; **P–JU**—1 sheet, No. 9855, microf. 727: III. 4).

"Frequent in dry fields near the sea."

Borreria verticillata *(L.) G. F. W. Mey.*—FWTA. ed. 2, 2: 220 (1963).

Spermacoce globosa Schum. et Thonn. in Schum. 1827: 73 and 1828: 93; FTA. 3: 240 (1877); Junghans 1962: 87.

Exsicc.: Ada, 1784, *Isert* s.n. (**C**—1 sheet, microf. 104: III. 2); R. Volta, *Thonning* 348 (**C**—3 sheets microf. 98: II. 6, 7; 104: III. 4).

"Frequently in the vicinity of the R. Volta."

"*Hedyotis verticillata* comes very near our plant, but according to Dillenius Hortus Elthamensis p. 369 both stems and leaves have stiff hairs whereas our plant is quite smooth."

[**Borreria sp.** ? *Spermacoce globosa* sensu Junghans 1962: 87.

Exsicc.: "Guinea?" "Schum." s.n. (**C**—1 sheet, microf. 105: I. 1, 2).

NOTE: I have been unable to match this plant exactly with a West African species, and perhaps its provenance is elsewhere than West Africa—F.N.H.]

Canthium horizontale (*Schumacher*) *Hiern*—FWTA. ed. 2, 2: 182 (1963).

Phallaria horizontalis Schumacher 1827: 112 and 1828: 132 (no MS.); Junghans 1962: 86.

Exsicc.: "Guinea", *Thonning* s.n. (**C**—1 sheet, type, microf. 79: I. 1–2); "Guinea" (**C**—1 sheet, microf. 79: I. 3; **P–JU**—1 sheet, No. 9797, microf. 724: II. 2).

Canthium multiflorum (*Thonning*) *Hiern*—FWTA. ed. 2, 2: 182 (1963).

Psychotria multiflora Thonning in Schum. 1827: 109 and 1828: 129; Junghans 1962: 86.

Exsicc.: Asiama, *Thonning* 300 (**C**—1 sheet, type, microf. 89: III. 1); "e Guinea" (**FI**—1 sheet; **P–JU**—1 sheet, No. 9830, microf. 726: II. 1).

"Found at Asiama, flowering in May."

NOTE: The Florence sheet is labelled "Psychotria umbellata".—F.N.H.

Chassalia kolly (*Schumacher*) *Hepper* in Kew Bull. 16: 330 (1962); FWTA. ed. 2, 2: 192 (1963).

Psychotria kolly Schumacher 1827: 110 and 1828: 130 (no MS); FTA. 3: 203 (1877); Junghans 1962: 86.

Exsicc.: "Guinea", *Thonning* s.n. (**C**—1 sheet, type, microf. 89: II. 7); "e Guinea" (**P–JU**—1 sheet, No. 9829, microf. 726: I. 7).

VERNACULAR NAME: 'Kolly-tjo'.

Cephaelis peduncularis *Salisb.*—FWTA. ed. 2, 2: 204 (1963).

Rubiaceae indet. in Junghans 1962: 87.

Exsicc.: Sawi, 1785, *Isert* s.n. (**C**—1 sheet, microf. 90: I. 6).

NOTE: This is a tentative determination.—F.N.H.

Cremaspora triflora (*Thonning*) *K. Schum.*—FWTA. ed. 2, 2: 148 (1963).
Cremaspora africana Benth.—FTA. 3: 126 (1877).

Psychotria triflora Thonning in Schum. 1827: 108 and 1828: 128; Junghans 1962: 87.

Exsicc.: Asiama, *Thonning* 299 (**C**—2 sheets, types, microf. 89: III. 5–7); "e Guinea" (**P–JU**—1 sheet, No. 9828, microf. 726: I. 6; **S**—1 sheet, type).

"Grows in the neighbourhood of Asiama."

Diodia sarmentosa *Sw.*—Verdcourt in FTEA. Rubiac. 1: ined. (1976). *D. scandens* Sw.—FWTA. ed. 2, 2: 216 (1963). *Spermacoce? pilosa* (Schum. et Thonn.) DC.—FTA. 3: 235 (1877).

D. pilosa Schum. et Thonn. in Schum. 1827: 76 and 1828: 96; Junghans 1962: 84.

Exsicc.: Aquapim, *Thonning* 212 (**C**—2 sheets, types, microf. 36: I. 4, 5); "e Guinea (**P–JU**—1 sheet, No. 9854, type, microf. 727: III. 4).

"Not frequent in valleys in Aquapim."

Diodia serrulata (*P. Beauv.*) *G. Tayl.*—FWTA. ed. 2, 2: 216 (1963).

D. maritima Thonning in Schum. 1827: 75 and 1828: 95; FTA. 3: 231 (1877); Junghans 1962: 84.

Exsicc.: *Thonning* 255 (no type material traced).

"Frequent on sandy coasts."

Gardenia ternifolia *Schum. et Thonn.* in Schum. 1827: 147 and 1828: 167; FWTA. ed. 2, 2: 123 (1963); Junghans 1962: 84.

Exsicc.: Gah and Adampi, *Thonning* 140 (**C**—2 sheets, types, microf. 43: III. 5–7; 44: I. 1–2; "e Guinea" (**FI**—1 sheet; **G–DC**—1 sheet, 4: 382. 26, type, microf. 686. III. 2).

Vernacular name: 'Paettaeplae-bi'.

"In the provinces Gah and Adampi; not common."
"*G. thunbergii* which comes near our plant, differs however in that the perianth opens at the side, has below the margin hood-shaped stalked lobes, and is larger, more swollen and smoother, also the corolla-tube is thicker and shorter and the collar-lobes are nearly three times as broad."

G. medicinalis Vahl ex Schum. 1827: 148 and 1828: 168; Junghans 1962: 84.

Exsicc.: Fida, 1786, *Isert* s.n. (**C**—1 sheet, type, microf. 43: III. 3–4).

"In woods in Fida."
"According to Isert the natives make use of it in different illnesses."

Geophila obvallata (*Schumacher*) *F. Didr.*—FWTA. ed. 2, 2: 206 (1963).

Psychotria obvallata Schumacher 1827: 111 and 1828: 131 (no MS); Junghans 1962: 86.

Exsicc.: Aquapim, 1786, *Isert* s.n. (**C**—2 sheets, types, microf. 89: III. 2–4).

"At Aquapim."

"Related to *P. repens, herbacea* and other species with herbaceous creeping
stem, but differs in the named characters. It was not observed by Thonning
but is found in Isert's herbarium."

Kohautia virgata (*Willd.*) *Bremek.*—FWTA. 2, 2: 209 (1963). *Oldenlandia
virgata* (Willd.) DC.—FTA. 3: 59 (1877).

Hedyotis virgata Willd. 1798a: 567 ("Habitat in Guinea"); Schum. 1827: 69
and 1828: 89; Junghans 1962: 85.

EXSICC.: Ada, 1784, *Isert* s.n. (**B**—1 sheet, No. 2596, type, microf. 173: I. 5;
C—1 sheet, type, microf. 50: II. 3); "Guinea", *Thonning* 178 (**C**—2 sheets,
microf. 50: II. 2, 4, 5; **S**—1 sheet); "e Guinea" (**C**—1 sheet, microf. 50:
II. 6, 7; **FI**—1 sheet; **P–JU**—1 sheet, No. 9851, microf. 727: III. 2).

"Occurs frequently in cultivated places; flowers in June."

Mitracarpus villosus (*Sw.*) *DC.*—Verdcourt in Kew Bull. 30: 317 (1975),
and in FTEA. Rubiac. 1: ined. (1976). *M. scaber* Zucc.—FTA. 3: 243 (1877);
FWTA. ed. 2, 2: 222 (1963).

EXSICC.: Whydah, 1785, *Isert* s.n. (**C**—1 sheet, microf. 98: II. 5).

Staurospermum verticillatum Schum. et Thonn. in Schum. 1827: 73 and 1828:
93; Junghans 1962: 87.

EXSICC.: "Guinea" coast, *Thonning* 242 (**C**—3 sheets, types, microf. 104: III.
3, 5–7); "e Guinea" (**P–JU**—1 sheet, No. 9853, microf. 727: III. 4 partly).

"Common near the seashore in the open fields."

Mitragyna inermis (*Willd.*) *Kuntze*—FWTA. ed. 2, 2: 161 (1963).
Uncaria inermis Willd. 1793: 199–201, Pl. 3 ("Habitat in Guinea"); Junghans
1962: 87.

Nauclea africana Willd. 1798a: 929; Schum. 1827: 104 and 1828: 124.
Mitragyne africana (Willd.) Korth—FTA. 3: 40 (1877).

EXSICC.: Ada, 1784, *Isert* s.n. (**B**—1 sheet, No. 3908, type, microf. 274:
II. 5; **C**—1 sheet, type, microf. 108: I. 6); "Guinea", *Thonning* 16 (**C**—1
sheet, microf. 108: II. 1, 2; **S**—1 sheet); "e Guinea" (**C**—1 sheet, microf.
108: II. 3, 4; **FI**—1 sheet; **P–JU**—1 sheet, No. 10015, microf. 738: II. 2).

VERNACULAR NAME: 'Kiná-tjo'.

"Common on the plains."
"The wood is hard, fine, yellowish with reddish veins and takes a very
handsome polish."

NOTE: One of the two Thonning sheets mentioned by Junghans is in fact
Conocarpus erectus.—F.N.H.

Morinda lucida *Benth.*—FWTA. ed. 2, 2: 189 (1963). *Morinda citrifolia*
sensu FTA. 3: 191 (1877), non L.

Rubiaceae indet. in Junghans 1962: 87.

Exsicc.: Ada, 1784, *Isert* s.n. (C—1 sheet, microf. 18: II. 2).

Psychotria? chrysorhiza Thonning in Schum. 1827: 111 and 1828: 131 (no MS No.); Junghans 1962: 86.

VERNACULAR NAME: 'Boj-tegi-tjo'.

Exsicc.: *Thonning* s.n. (no type specimens traced).

Mussaenda elegans *Schum. et Thonn.* in Schum. 1828: 117 and 1828: 137; DC. 1830: 372; FWTA. ed. 2, 2: 167 (1963); Junghans 1962: 85.

Exsicc.: Aquapim, *Thonning* 170 (no type material at C; G–DC—1 sheet, 4: 372. 19, type, microf. 685: I. 4; S—1 sheet, type).

"On the mountains of Aquapim, and in the neighbouring valleys. Flowers in May."

Mussaenda erythrophylla *Schum. et Thonn.* in Schum. 1827: 116 and 1828: 136; FWTA. ed. 2, 2: 165, fig. 234 (1963); Junghans 1962: 85.

Exsicc.: Aquapim mountains, *Thonning* 93 (C—1 sheet, type, microf. 69: III. 3; G–DC—1 sheet, 4: 371. 8, type, microf. 684: II. 6).

"On the mountains of Aquapim, more rarely in the neighbouring valleys. Flowers in May and Sept."

Mussaenda isertiana *DC.* 1830: 371 ("in Guineâ legit Isert")—FTA. 3: 67 (1877); FWTA. ed. 2, 2: 167 (1963).

M. macrophylla Vahl ex Schum. in Schum. 1827: 118 and 1828: 138; Junghans 1962: 86, non Wallich (1824).

Exsicc.: Whydah, 1785, *Isert* s.n. (C—1 sheet, type, microf. 69: III. 4).

"Near Whydah according to Isert. Found in Isert's collection."

Nauclea latifolia *Sm.*—FWTA. ed. 2, 2: 163 (1963). *Sarcocephalus esculentus* Afzel. ex Sabine—FTA. 3: 381 (1877).

Cephalina esculenta (Afzel. ex Sabine) Schum. et Thonn. in Schum. 1827: 105 and 1828: 125; Junghans 1962: 84.

Exsicc.: "Guinea", *Thonning* 84 (C—2 sheets, type, microf. 18: I. 5–7).

VERNACULAR NAME: 'Baabylaa-najrié'.

"Here and there amongst bushes; flowers in May and June."
"The natives eat the ripe fruit, which is slightly acid. Th."

Oldenlandia corymbosa *L.*—FTA. 3: 62 (1877); FWTA. ed. 2, 2: 211 (1963).

Hedyotis longifolia Schum. et Thonn. in Schum. 1827: 70 and 1828: 90; Junghans 1962: 85. *Oldenlandia longifolia* (Schum. et Thonn.) DC. 1830: 426.

Exsicc.: "Guinea", *Thonning* 177 (C—3 sheets, types, microf. 49: III. 1–3);

"e Guinea" (**P–JU**—1 sheet, No. 9857, microf. 727: III. 7; **S**—1 sheet, labelled No. 41).

"Not frequent."

"*H. biflora* Horn. hort. Hafn. p. 130 differs from our plant in its more delicate stems, shorter internodes, narrower leaves which are not an inch long and the peduncles which are many times shorter, 3–4-flowered in cultivated place."

NOTE: Both the Thonning sheets are erroneously labelled by Bremekamp: one as *O. linearis* and the other as *O. lancifolia*. These errors led Bremekamp (1952: 230, discussed on p. 234) into supposing that *Hedyotis longifolia* and *H. lancifolia* were synonyms.—F.N.H.

Oldenlandia lancifolia (*Schumacher*) *DC.* 1830: 425; Bremek. 1952: 230–234; FWTA. ed. 2, 2: 212 (1963); Junghans 1962: 84.

Hedyotis lancifolia Schumacher 1827: 72 and 1828: 92.

EXSICC.: Aquapim, *Thonning* 210 (**C**—2 sheets, microf. 49: II. 5, 6—designated here lectotype—7; **S**—1 sheet); "e Guinea" (**C**—1 sheet, microf. 49: III. 4, 5; **P–JU**—1 sheet, No. 9869, microf. 728: II. 7).

"In damp places in the valleys of Aquapim."

NOTE: The "e Guinea" sheet was correctly determined by Bremekamp, but it was included with the last species in Junghans list.—F.N.H.

Oxyanthus racemosus (*Schum. et Thonn.*) *Keay*—FWTA. ed. 2, 2: 129 (1963). *Oxyanthus tubiflorus* sensu FTA. 3: 107 (1877), non (Andr.) DC.

Ucriana racemosa Schum. et Thonn. in Schum. 1827: 107 and 1828: 127; Junghans 1962: 87.

EXSICC.: Begusso, *Thonning* 286 (**C**—3 sheets, types, microf. 107: III. 7; 108: I. 1–5); "e Guinea" (**C**—1 sheet, microf. 107: III. 6; **P–JU**—1 sheet, No. 9934, microf. 733: I. 3).

"Rare; in some few places in Aquapim e.g. Begusso."

Pavetta corymbosa (*DC.*) *F. N. Williams*—FWTA. ed. 2, 2: 138 (1963). *Pavetta baconia* Hiern—FTA. 3: 176 (1877).

Ixora nitida Schum. et Thonn. in Schum. 1827: 77 and 1828: 97; Junghans 1962: 85.

EXSICC.: Whydah, 1785, *Isert* s.n. (**C**—2 sheets, microf. 77: III. 2–4); Asiama and Dadintam, *Thonning* 271 (**C**—1 sheet, type, microf. 61: II. 1); "e Guinea" (**P–JU**—1 sheet, No. 9778, microf. 723: I. 1).

VERNACULAR NAME: 'Koi-tio'.

"Grows frequently at Asiama and Dadintam. Flowers in April."

"As this plant in habit agrees so entirely with the *Ixora* species, Thonning has referred it to this genus, although he has not seen the fruit."

Pavetta genipifolia *Schumacher* 1827: 78 and 1828: 98 (no MS.); FWTA. ed. 2, 2: 139 (1963); Junghans 1962: 86. *P. megistocalyx* K. Krause—Bremek. in Fedde, Repert. 37: 136 (1934).

Exsicc.: Aquapim, 1786, *Isert* s.n. (**C**—1 sheet, probably type, microf. 78: II. 3); "e Guinea" (**P–JU**—1 sheet, No. 9788, microf. 723: II. 7).

Note: If the **P–JU** sheet is an Isert duplicate, then it is unusual since Vahl distributed Thonning's specimens to Paris—F.N.H.

Pavetta subglabra *Schumacher* 1827: 78 and 1828: 98 (no MS.); FWTA. ed. 2, 2: 140 (1963); Junghans 1962: 86.

Exsicc.: "Guinea", *Thonning* s.n. (**C**—2 sheets, types, microf. 78: II. 4, 5; **S**—2 sheets, types); "e Guinea" (**P–JU**—1 sheet, No. 9789, microf. 723: III. 1).

"*P. indica* differs from our plant in that it is quite smooth. *P. longiflora* in its lanceolate-elliptic leaves, more contracted at the base, longer and smooth leaf-stalks, and in that the tube of the corolla is 9 inches long and thicker, style much shorter, the corymbs few-flowered. S."

Pentodon pentandrus (*Schum. et Thonn.*) *Vatke*—FWTA. ed. 2, 2: 213 (1963). *Oldenlandia macrophylla* DC.—FTA. 3: 63 (1877).

Hedyotis pentandra Schum. et Thonn. in Scum. 1827: 71 and 1828: 91; Junghans 1962: 85.

Exsicc.: Atocco, 1784, *Isert* s.n. (**C**—1 sheet, microf. 49: III. 6); Aquapim, *Thonning* 211 (**C**—5 sheets, types, microf. 49: III. 7; 50: I. 1–7; II. 1); "e Guinea" (**P–JU**—1 sheet, No. 9868, microf. 728: II. 7; **S**—1 sheet, type).

"The valleys at Aquapim."
"Perhaps the same as *H. racemosa* Lamarck, Encycl. 3 p. 76. Tab. 62, f. 2, although the leaves are admitted to be ovate and acute on both sides; the author does not state the number of stamens. Rheed. Hort. X tab. 25 agrees somewhat with our plant, but I do not venture to declare it as the same."

Pouchetia africana *DC.*—FWTA. ed. 2, 2: 132 (1963).

Rubiaceae indet. in Junghans 1962: 87.

Exsicc.: "Guinea, Hb. Vahl ded. L'Héritier" s.n. (**C**—1 sheet, microf. 89: I. 5, 6).

Note: No doubt this is a Thonning specimen.—F.N.H.

Psychotria calva *Hiern*—FWTA. ed. 2, 2: 200 (1963).

P. umbellata Thonning in Schum. 1827: 109 and 1828: 129; Junghans 1962: 87.

Exsicc.: Adah, *Thonning* 322 (**C**—3 sheets, types, microf. 90: I. 1–5).

"Grows at Adah; not common."

Rubiaceae indet. in Junghans 1962: 87.

EXSICC.: Fida, 1784, *Isert* s.n. (**C**—1 sheet, microf. 94: I. 3, 4).

Rutidea parviflora *DC.*—FWTA. ed. 2, 2: 146 (1963).

Rubiaceae indet. in Junghans 1962: 87.

EXSICC.: "Guinea, Hb. Vahl ded. L'Héritier" s.n. (**C**—1 sheet, microf. 44: I. 3, 4).

NOTE: No doubt this is a Thonning specimen.—F.N.H.

Vangueriopsis spinosa *(Schum. et Thonn.) Hepper* in Kew Bull. 17: 170 (1963); FWTA. ed. 2, 2: 179 (1963).

Phallaria spinosa Schum. et Thonn. in Schum. 1827: 113 and 1828: 133.

EXSICC.: "Guinea", *Thonning* 287 (**P–JU**—1 sheet, No. 9798, type, microf. 724: II. 3).

"Not common."

"The name is chosen because the stigma resembles the pileus of *Phallus impudicus*."

NOTE: Although no type has been traced in Copenhagen, the sheet in Paris must be a duplicate of *Thonning* 287.—F.N.H.

Virectaria procumbens *(Sm.) Bremek.*—FWTA. ed. 2, 2: 208 (1963).

Rubiaceae indet. in Junghans 1962: 87.

EXSICC.: Aquapim, 1786, *Isert* s.n. (**C**—1 sheet, microf. 109: II. 6).

RUTACEAE

Afraegle paniculata *(Schum. et Thonn.) Engl.*—FWTA. ed. 2, 1: 687 (1958).

Citrus paniculata Schum. et Thonn. in Schum. 1827: 378 and 1829: 152; Junghans 1962: 87.

EXSICC.: "Guinea", *Thonning* 291 (**C**—2 sheets, types, microf. 20: III. 2–5); "e Guinea" (**C**—1 sheet, microf. 20: III. 6–7; **P–JU**—1 sheet, No. 11933, microf. 884: III. 4).

VERNACULAR NAME: 'Koklo-tjo'.

"The fruit is collected by the natives and burnt to an ash from which a lye is extracted which is boiled with palm oil to make soap. Th."

[Amyris sylvatica *Jacq.*

EXSICC.: "e Guinea?" (**C**—1 sheet, microf. 1: II. 3, 4).

NOTE: Although included in the microfiche, this species is native in Central America and the specimen is unlikely to have been collected in West Africa. See note on p. 141—F.N.H.]

Citropsis articulata (*Willd. ex Spreng.*) *Swingle et Kellermann*—FWTA. ed. 2, 1 : 688 (1958).

Citrus articulata Willd. ex Spreng. 1826: 334 ("Guinea. *Isert*"); Junghans 1962: 87.

EXSICC.: "Guinea", *Isert* (**B**—1 sheet, No. 14357, type, microf. 1027: II. 7; no type material at **C**).

Clausena anisata (*Willd.*) *Hook. f. ex Benth.*—FWTA. ed. 2, 1 : 686 (1958).

Amyris anisata Willd. 1799: 337 ("Habitat in Guinea"); Schum. 1827: 191 and 1828: 211; Junghans 1962: 87.

EXSICC.: *Isert* s.n. (**B**—1 sheet, No. 7292, type, microf. 502: II. 5; **C**—1 sheet type?, microf. 5: III. 5, 6); Christiansborg, 1784, *Isert* s.n. (**C**—1 sheet, type?, microf. 6: I. 3); Accra, 1785, *Isert* s.n. (**C**—1 sheet, type?, microf. 5: III. 7; 6. I. 1, 2); "Guinea", *Thonning* 29 (**C**—1 sheet, microf. 5: III. 3, 4); "e Guinea" (**P–JU**—1 sheet, No. 15973, microf. 1159: III. 7).

VERNACULAR NAME: 'Abami-tio' or 'Abami-aulage-tio'; 'Anistree' (European).

"Grows here and there. Rather common."
"The whole plant has an anise-like odour and taste; especially the leaves, flowers and seed; the root less and the wood least. The leaves boiled in water are commonly used against internal sickness, in a warm bath; in all inflammatory illnesses the bath is harmful. An infusion of the root is drunk in various stomach disorders. For facial swellings (Aboa in the Akkra language) the root is finely macerated with lemon juice and grains-of-Paradise [*Aframomum*], and smeared over the face."

Fagara zanthoxyloides *Lam.*—Willd. 1798a: 667 (as to the plant from Guinea); FWTA. ed. 2, 1: 685 (1958); Junghans 1962: 88.

EXSICC.: "Guinea", *Isert* s.n. (**B**—1 sheet, No. 3046, microf. 204: II. 1).

Zanthoxylum polygamum Schum. et Thonn. in Schum. 1827: 433 and 1829: 207.

EXSICC.: "Guinea", *Isert* s.n. (**C**—1 sheet, microf. 111: II. 6, 7); "Guinea", *Thonning* 46 (**C**—6 sheets, types, microf. 111: II. 5; III. 1–6; 112: I. 1–4); "e Guinea" (**P–JU**—1 sheet, No. 12991C, microf. 959: I. 2).

VERNACULAR NAME: 'Hah-tio'.

"Frequent."
"Near the coast, where the sea wind subjugates most plants with its destructive fog from the sea-water which it bears from the surf, it never reaches its natural size but remains as a shrub; the farther from the shore the better it thrives; 3–4 miles inland trees occur the size of average oaks; whose trunks also resemble them in figure and extent. The wood has various features which make it serviceable for furniture-work; it is strong, hard, heavy, a more beautiful yellow in colour than box, but like this it later loses its yellow colour; it is as fine as the average kind of mahogany; it contains much resin, which is so well distributed that it does not exude from the tree when it is worked as timber, but on the contrary causes it to assume a shiny

and fine polish; I have never found it attacked by worms; it is frequently interspersed with knots. As an average size of planks which this tree is capable of yielding, one may assume 5 to 8 inches in breadth and 2 ells in length; selected trees may perhaps yield planks of 12 to 16 inches in breadth but hardly more than 2 ells in length, as the trunk is generally twisted and curved. The bark of the root is used by the natives to expel gout-pains; along with grains-of-Paradise [*Aframomum*], it is finely grated and rubbed on the painful place. From the wood, especially the knotty kind, torches are made which are generally used for lighting amongst the poor people. From the prickles on ageing trees bullet moulds are made with carved ornaments for printing with colour on the body. In toothache the finely ground root bark is rubbed in externally on the cheek and a decoction is now and then applied to the teeth. Th."

SAPINDACEAE

Allophylus africanus *P. Beauv.*—FWTA. ed. 2, 1: 713 (1958). *Schmidelia africana* (P. Beauv.) DC.—FTA. 1: 421 (1868).

Ornithrophe tristachyos Schum. et Thonn. in Schum. 1827: 188 and 1828: 208; Junghans 1962: 89.

EXSICC.: Aquapim, *Thonning* 229 (**C**—5 sheets, types, microf. 74: II. 5–7, III. 1–7).

VERNACULAR NAME: 'Tatadua'.

"Amongst other shrubs on formerly cultivated places and in the neighbourhood of Aquapim."

Allophylus spicatus (*Poir.*) *Radlk.*—FWTA. ed. 2, 1: 714 (1958). *Ornithrophe spicata* Poir.—Junghans 1962: 89.

Ornithrophe magica Schum. et Thonn. in Schum. 1827: 186 and 1828: 206. *Schmidelia magica* (Schum. et Thonn.) Bak.—FTA. 1: 423 (1868).

EXSICC.: "Guinea", *Thonning* 57 (**C**—3 sheets, types, microf. 73: III. 4–7; 74: I. 1–2); "e Guinea" (**C**—2 sheets, microf. 73: III. 2, 3; 74: I. 3, 4; **FI**—1 sheet; **P–JU**—1 sheet, No. 11373, microf. 842: I. 6).

VERNACULAR NAME: 'Tadadua'.

"Grows here and there, not rare."
"The bark of the root is sometimes used for application to old leg injuries. The tree and leaves are often used as a fetish in cases of illness; a branch $1\frac{1}{2}$ ells in length is cut off, the bark is scraped off, smeared over with white clay-soil, one end is enwound with bast-rope which is smeared with a red kind of soil; the other end is inserted in the earth in a definite place ordered by the fetish priest or the doctor, usually at a cross road, often an egg or other trifle is laid with it and often several such sticks are set from time to time when the illness will not give way. The leaves are put into cold water with which the sick person is washed, after which the water is thrown out on the ground at the inserted stick. Th."

Anacardiaceae indet. in Junghans 1961: 322.

Exsicc.: Whydah, 1784, *Isert* s.n. (**C**—1 sheet, microf. 6: I. 4).

Blighia sapida *König*—FTA. 1: 426 (1868); FWTA. ed. 2, 1: 722 (1958).

Cupania edulis Schum. et Thonn. in Schum. 1827: 190 and 1828: 210; Junghans 1962: 88.

Exsicc.: "Guinea", *Isert* "54" (**C**—1 sheet, microf. 28: III. 7); Christiansborg, *Thonning* 150 (**C**—3 sheets, types, microf. 28: III. 5, 6; 29: I. 1–4).

Vernacular name: 'Atia-tjo'; 'Vild kaschu' (= Wild cashew) (European).

"Grows in the fields to the north of Christiansborg and Fredensborg. Flowers in November and bears fruit in January."

"The tree is almost the size of an average oak. The receptaculum seminis is eaten as fruit, but one must carefully separate the funiculi umbilicales from it, as this latter, according to what the natives say, is a very powerful poison. The bark is macerated with lemon juice and is used as a poultice, or is merely rubbed in on swollen testicles. Th."

Cardiospermum grandiflorum *Sw.*—FWTA. ed. 2, 1: 711 (1958).

C. hirsutum Willd. 1799: 467 ("Habitat in Guinea"); Schum. 1827: 196 and 1828: 216.

Exsicc.: "Guinea", *Isert* s.n. (**B**—1 sheet, No. 7732, type, microf. 533: II. 1); Aquapim, *Thonning* 155 (**C**—1 sheet, microf. 14: III. 3, 4).

Vernacular name: 'Sablabo'.

"Here and there but chiefly near Aquapim, in copse-woods; it flowers most times of the year.'

"It is used as a fetish. Th."

"*C. grandiflorum* which much resembles our plant, differs in its smaller stem, smaller globular smooth capsules, and less hairy leaves. S."

Cardiospermum halicacabum *L.*—FTA. 1: 417 (1868); FWTA. ed. 2, 1: 711 (1958).

C. glabrum Schum. et Thonn. in Schum. 1827: 197 and 1828: 217; Junghans 1962: 88.

Exsicc.: "Guinea", *Thonning* 122 (**C**—2 sheets, types, microf. 14: II. 6, 7; III. 1, 2).

Vernacular name: 'Sablabó'.

"Grows here and there, preferably in the towns."

"Much like *C. halicacabum* but in this the leaves are smaller, and not so deeply incised, the outer leaf often pinnate or the outermost leaflet 3-lobed; the flower-stalks both the common and the individual ones are much shorter, and the capsules obcordate downy-haired. S."

"Used as a fetish, but not so frequently as the foregoing [*C. grandiflorum*]. Th."

Deinbollia pinnata (*Poir.*) *Schum. et Thonn.* in Schum. 1827: 242 and 1829:

16; FWTA. ed. 2, 1: 715 (1958); Junghans 1962: 88. *Ornitrophe pinnata*
Poir. 1808: 266. *Schmidelia pinnata* (Poir.) DC. 1824: 611. *Prostea pinnata*
(Poir.) Camb. 1829: 25, 39, pl. 1 fig. C.

Exsicc.: "Guinea", *Thonning* 248 (**C**—5 sheets, types, microf. 35: I. 7; II. 1,
3–7; III. 1–3; **FI**—1 sheet, type); "e Guinea" (**C**—3 sheets, microf. 35: I.
5, 6; II. 2; **P–JU**—1 sheet, No. 11388, microf. 843: III. 1).

Vernacular name: 'Badimanoplaa'.

"Here and there."

"The berries are tasteless; children who eat them from curiosity get sore
lips. Th."

"Named after Provost Deinboll, known for his research on the plants of
Finmark."

Dodonaea viscosa *Jacq.*—FWTA. ed. 2, 1: 724 (1958).

D. repanda Schum. et Thonn. in Schum. 1827: 194 and 1828: 214; FTA. 1:
433 (1868); Junghans 1962: 88.

Exsicc.: Pottebra, 1784, *Isert* "240" (**C**—1 sheet, microf. 37: I. 1); Volta R.,
Thonning 235 (**C**—2 sheets, types, microf. 37: I. 2–4).

"Common on the banks of the Volta."

"*D. viscosa* much resembles the plant, but the leaves are much longer,
narrower, almost cuneate, entire-margined (not emarginate above) and all
viscous, the lateral veins more across; the ripe fruits are also larger and darker.
In *D. latifolia* the leaves are oblong-ovate, entire-margined, rounded and
provided with a short point, nearly 4 inches long, 1½ inches broad. The
racemes are situated in the leaf-axils and terminal and are compound. S."

Pancovia bijuga *Willd.* 1799: 285 ("Habitat in Guinea"); FWTA. ed. 2,
1: 718 (1958); Junghans 1962: 89.

Exsicc.: "Guinea", *Isert* s.n. (**C**—1 sheet, type, microf. 75: I. 2, 3; **B**—2
sheets, No. 7126, types, microf. 491: II. 5, 6—see note).

Note: This microfiche has been incorrectly numbered: 491 should be 492
and vice versa—F.N.H.

Paullinia pinnata *L.*—FTA. 1: 419 (1868); FWTA. ed. 2, 1: 710 (1958).

P. uvata Schum. et Thonn. in Schum. 1827: 195 and 1828: 215; Junghans
1962: 89.

Exsicc.: "Guinea", 1784, *Isert* s.n. (**C**—1 sheet, microf. 78: I. 2); Aquapim,
Thonning 156 (**C**—5 sheets, types, microf. 78: I. 1, 3–7; II. 1, 2); "e Guinea"
(**P–JU**—1 sheet, No. 11353C, microf. 841: I. 3).

"Here and there in the thickets, but chiefly in and in the neighbourhood
of Aquapim. Flowering for the greater part of the year."

"In *P. pinnata* which comes nearest, the leaflets are oblong-ovate, acute
or nearly acuminate, longer and narrower; the racemes longer, the flowers
in clusters, clusters larger, farther from each other so that they do not form a
cylindric spike; the calices are smooth and not covered with silk-downy hairs

under the lens; the capsules pear-shaped longer, three angled, more contracted at the base."

Aphania senegalensis (*Juss. ex Poir.*) *Radlk.*—FWTA. ed. 2, 1: 716
(1958).

Ornithrophe thyrsoides Schum. et Thonn. in Schum. 1827: 185 and 1828:
205; Junghans 1962: 89. *Schmidelia thyrsoides* Baker—FTA. 1: 423 (1868).

EXSICC.: "Guinea", *Isert* "150" (**C**—1 sheet, microf. 74: I. 5); "Quitta",
Thonning 342 (**C**—3 sheets, types, microf. 74: I. 6–7, II. 1–4).

"At Quitta."

SAPOTACEAE

Chrysophyllum albidum *G. Don*—FWTA. ed. 2, 2: 27 (1963).

Achras sericea Schumacher 1827: 179 and 1828: 99 (no MS.); FTA. 3: 501
(1877); Junghans 1962: 90.

EXSICC.: "Guinea", *Thonning* s.n. (**C**—2 sheets, type, microf. 2: I. 3–6).

"Thonning only brought leaves of this plant; they much resemble the
leaves of *A. macrophylla*; but these are smooth and shiny on both sides. S."

Synsepalum dulcificum (*Schum. et Thonn.*) *Daniell*—FWTA. ed. 2, 2: 22
(1963).

Bumelia dulcifica Schum. et Thonn. in Schum. 1827: 130 and 1828: 150;
Junghans 1962: 90.

EXSICC.: Aquapim, *Thonning* 168 (**C**—4 sheets, types, microf. 11: III. 4–7;
12: I. 1–4); "e Guinea" (**P–JU**—1 sheet, No. 7235, type, microf. 532: I. 5).

VERNACULAR NAME: 'Tahmi'.

"Not common in Aquapim."
"The berries are only slightly pulpy and have hardly any taste, but the
special peculiarity of betraying the taste so that after having eaten two or
three berries, everything tastes sweet, so that a lemon tastes like an orange,
vinegar like sweet wine etc. If one eats about a score of berries in the morning
the taste is retained in the mouth for nearly the whole day. As palm wine
(toddy) ferments so quickly that it can only be brought with difficulty about
5 miles from Aquapim to the coast towns without becoming sour, the wine
merchants usually bring with them some of these berries, whereby the
wine lovers at first please the palate and then intoxication from the sour
wine follows".

SAXIFRAGACEAE

Vahlia dichotoma (*Murr.*) *Kuntze*—FWTA. ed. 2, 1: 120 (1954); FTEA.
Vahliac. 2 (1975). *Bistella dichotoma* (Murr.) Bullock in Acta Bot. Neerl. 15: 85
(1966).

Oldenlandia pentandra Retz.—Schum. 1827: 83 and 1828: 103; FTA. 3: 65 (1877); Junghans 1962: 86.

Exsicc.: "Guinea", *Thonning* 236 (**C**—2 sheets, microf. 72: III. 4–6).

"Here and there."

"The Guinea plant differs from the Indian in its stronger thicker stem which is a foot in height; in the more swollen joints, segments 2 inches in length, and in the leaves being narrower and longer and the flower-stalks of the length of the leaves, an inch or over."

SCROPHULARIACEAE

Alectra vogelii *Benth.*—FWTA. ed. 2, 2: 368 (1963).

Scrophulariaceae indet. in Junghans 1962: 90.

Exsicc.: Whydah, 1785, *Isert* s.n. (**C**—1 sheet, microf. 96: I. 7).

Artanema longifolium (*L.*) *Vatke*—FTA. 4, 2: 327 (1906); FWTA. ed. 2, 2: 361 (1963).

Achimenes sesamoides Vahl—Schum. 1827: 281 and 1829: 55; Junghans 1962: 90.

Exsicc.: Dudun, *Thonning* 226 (no material traced).

"At Dudun".

Bacopa crenata (*P. Beauv.*) *Hepper* in Kew Bull. 14: 407 (1960); FWTA. ed. 2, 2: 359 (1963).

Erinus africanus L.—Schum. 1827: 278 and 1829: 52, as to the plant from Guinea; Junghans 1962: 90.

Exsicc.: Ningo, *Thonning* 334 (no material so labelled at **C**, but see note below).

"Grows at Ningo."

Herpestis thonningii Benth. in Hook. 1836: 58. *Moniera calycina* Hiern var. *thonningii* (Benth.) Benth. —FTA. 4, 2: 321 (1906).

Exsicc.: "Guinea", *Thonning* (334?) (**C**—1 sheet, microf. 65: I. 1, 2).

Note: This sheet apparently was not indexed by Junghans although it bears the name "Lindernia?", perhaps it is the missing No. 334—F.N.H.

Lindernia diffusa (*L.*) *Wettst.* var. **diffusa**—FWTA. ed. 2, 2: 364 (1963).

Scrophulariaceae indet. in Junghans 1962: 90.

Exsicc.: Whydah, 1784, *Isert* s.n. (**C**—1 sheet, microf. 109: I. 5).

Micrargeria filiformis (*Schum. et Thonn.*) *Hutch. et Dalz.*—FWTA. ed. 2, 2: 366 (1963). *M. scopiformis* (Klotzsch) Engl.—FTA. 4, 2: 457 (1906).

Gerardia filiformis Schum. et Thonn. in Schum. 1827: 272 and 1829: 46; Junghans 1962: 90.

Exsicc.: Prampram and Ningo, *Thonning* 366 (**C**—2 sheets, types, microf. 44: I. 5–7, II. 1; **P–JU**—1 sheet, No. 6097B, type, microf. 436: II. 5).

"In dry fields near Prampram and Ningo."

"Without more detailed investigation one could easily take this plant for *Gerardia tenuifolia* but in the latter the leaves are shorter, smooth, the flower-stalks filiform an inch long or a little shorter, and the calyces five-toothed. In *G. purpurea* the leaves are linear, shorter, and the calyces five-partite."

Scoparia dulcis *L.*—Schum. 1827: 79 and 1828: 99; FWTA. ed. 2, 2: 356 (1963); Junghans 1962: 90.

Exsicc.: "Guinea", *Thonning* 111 (**C**—1 sheet, microf. 98: I. 5, 6).

Vernacular name: 'Sjaa-blaa'.

"Fairly common."

"The American plant differs with a more delicate root, more branched stem, less fleshy broader leaves, and in that nearly all the flower-stalks are situated in pairs in the leaf-axils."

Striga aspera *(Willd.) Benth.*—FWTA. ed. 2, 2: 372 (1963).

Euphrasia aspera Willd. 1800b: 197 ("Habitat in Guinea"); Junghans 1962: 90.

Exsicc.: Ada, 1784, *Isert* s.n. (**B**—1 sheet, No. 11182, type, microf. 796: I. 4; **C**—1 sheet, type, microf. 11: II. 7).

Buchnera aspera (Willd.) Schum. 1827: 280 and 1829: 54.

Exsicc.: Ningo, *Thonning* 335 (**C**—1 sheet, microf. 11: II. 5, 6).

"Rare; found in an open dry field near Ningo in August."

"Although Willdenow's description is somewhat divergent, I nevertheless have no doubt that it is the same plant, as the specimen in Isert's herbarium entirely answers to our plant. For the rest it comes near *B. asiatica* and *B. gesneroides* from which however it adequately differs in the characters cited. As I have not seen the ripe capsule I dare not determine whether this species, with 3 bracts by the flowers, should rather perhaps be referred to *Piripea* Aubl. et Juss. p. 100. S."

Striga linearifolia *(Schum. et Thonn.) Hepper* in Kew Bull. 14: 416 (1960); FWTA. ed. 2, 2: 372 (1963).

Buchnera linearifolia Schum. et Thonn. in Schum. 1827: 279 and 1829: 53; Junghans 1962: 90.

Exsicc.: Fredriksberg, *Thonning* 284 (**C**—2 sheets, types, microf. 11: III. 1–3; **P–JU**—1 sheet, No. 5960, type, microf. 425: II. 5).

"Grows in a valley near Frederiksberg."

"Resembles *Buchnera euphrasioides* Vahl, but differs from it in its undivided stem, and therein that the uppermost leaves are more pointed and longer than the calyx and the lobes of the lower lip of the corolla are linear and pointed. S."

SOLANACEAE

Capsicum frutescens *L.*—FWTA. ed. 2, 2: 328 (1963).

Solanaceae indet. in Junghans 1962: 91.

EXSICC.: Whydah, 1785, *Isert* s.n. (**C**—2 sheets, microf. 14: II. 5, only 1 sheet photographed).

NOTE: The second sheet has turned up since Junghans prepared his enumeration.—F.N.H.

Datura metel *L.*—FWTA. ed. 2, 2: 326 (1963).

D. fastuosa L.—Schum. 1827: 118 and 1828: 138; Junghans 1962: 90.

EXSICC.: Fredriksberg, *Thonning* 305 (**C**—1 sheet, microf. 35: I. 3, 4).

"Rare; Thonning found only some few plants in the cotton plantation, Fredriksberg."

"Our plant differs from the Indian plant and similarly from the cultivated one, in that the leaves are more angular with more obtuse angles and less hairy; the calyx is longer and narrower, longer than the corolla tube."

"The natives, owing to its rarity were unaware of its poisonous properties."

Physalis angulata *L.*—Schum. 1827: 120 and 1828: 140; FWTA. ed. 2, 2: 329 (1963); Junghans 1962: 91.

EXSICC.: "Guinea", *Thonning* 109 (**C**—2 sheets, microf. 84: III. 5–7).

VERNACULAR NAME: 'Amotobi'.

"Here and there, preferably in shady places in good soil."

"Adampi girls, who without having observed the religious ceremonies, have become pregnant, try to expel the foetus with a decoction of this plant, which they partly drink, partly wash the generative parts with it, partly use as a laxative. The leaves are also used against an itchy eruption and against mumps; they are crushed and the body is rubbed with them."

Schwenckia americana *L.*—FWTA. ed. 2, 2: 327 (1963).

S. guineensis Schum. et Thonn. in Schum. 1827: 8 and 1828: 28; Junghans 1962: 91.

EXSICC.: "Guinea", *Thonning* 114 (**C**—2 sheets, type, microf. 96: II. 1–3; **P–JU**—1 sheet, No. 6261B, type, microf. 449: II. 7).

"Common on roads or open and dry fields, at all seasons, but chiefly at rainy periods."

"*S. americana* in Vahl's herbarium differs from our plant in that the flowers are long-stalked, tube of corolla 3 times as long as the calyx, narrower and cylindric, the capsules smaller, the leaves smaller, narrower and more pointed, the branches less downy-haired. S."

Solanum anomalum *Thonning* in Schum. 1827: 126 and 1828: 146; FWTA. ed. 2, 2: 334 (1963); Junghans 1962: 91.

Exsicc.: "Guinea", *Thonning* 135 (**C**—5 sheets, types, microf. 101: I. 4; II. 2–7; III. 1, 2; **G–DC**—1 sheet, 13, 1: 259. 617, type, microf. 2081: II. 8); "e Guinea" (**C**—2 sheets, microf. 101: I. 5–7; II. 1; **FI**—1 sheet; **P–JU**—1 sheet, No. 6431, microf. 465: III. 7).

VERNACULAR NAME: 'Asogagaplae' or 'Sissa-sussoa'.

"Here and there, although not common."
"The natives use the juice of the berries to smear on sores of the ear."

Solanum dasyphyllum *Schum. et Thonn.* in Schum. 1827: 126 and 1828: 146; FWTA. ed. 2, 2: 334 (1963); Junghans 1962: 91.

Exsicc.: "Guinea", *Thonning* 144 (**C**—2 sheets, types, microf. 102: I. 1–4; **G–DC**—1 sheet, 13, 1: 313. 731, type, microf. 2085: III. 7).

VERNACULAR NAME: 'Atropo-bah'.

"Is cultivated by the natives."
"The unripe fruit is used in the same way as *S. oleraceum*."

Solanum gilo *Raddi* var. **gilo**—FWTA. ed. 2, 2: 332 (1963).

S. geminifolium Thonning in Schum. 1827: 121 and 1828: 141; FTA. 4, 2: 223 (1906); Junghans 1962: 91.

Exsicc.: "Guinea", *Thonning* 143 (**C**—1 sheet, type, microf. 102: II. 7).

VERNACULAR NAME: 'Sebae'.

"Is cultivated."
"The fruit is prepared in different ways; it is chiefly used in soup or is prepared as a mash with salt, Spanish pepper, palm oil and some dry fish."

Solanum indicum *L.* subsp. **distichum** *(Schum. et Thonn.) Bitter*—FWTA. ed. 2, 2: 333 (1963).

S. distichum Schum. et Thonn. in Schum. 1827: 122 and 1828: 142; FTA. 4, 2: 223 (1906); Junghans 1962: 91.

Exsicc.: Fida, 1785, *Isert* s.n. (**C**—1 sheet, microf. 102: I. 6); Aquapim, *Thonning* 199 (**C**—3 sheets, types, microf. 102: I. 5, 7; II. 1–3).

VERNACULAR NAME: 'Sussoa'.

"In Aquapim."
"The berries are eaten by children, without harm."

Solanum macrocarpon *L.*—FWTA. ed. 2, 2: 334 (1963). *S. thonningianum* Jacq. 1811–16: 123–125, t. 83; Junghans 1962: 91.

S. atropo Schum. et Thonn. in Schum. 1827: 124 and 1828: 144.

Exsicc.: "Guinea", *Thonning* 117 (**C**—2 sheets, types, microf. 101: III. 3–6; **LE**—1 sheet, type).

Vernacular names: 'Koa-fyé' (plant); 'Atropo' (fruit), 'Fankvau' (cooked leaves).

"Is cultivated."
"Wonder if it is the same plant as *S. thonningianum* Jacq. Eclog. plant. rar.?"
"The unripe fruit is boiled in soup or is prepared with half-rotten dried fish, Spanish pepper, salt, palm oil and onions, a dish which the natives regard as a delicacy. The leaves yield a very good cabbage which is stewed in the same way as the fruit; this dish is called 'Fankvau' and is one of the native's choicest foods."

Note: Junghans 1962: 91 observes that "the MS shows that the plant was first regarded as belonging to *S. macrocarpon* L. and later was called *S. oleraceum* and *S. atropo*". According to Jacquin plants were grown at Vienna University Garden from seeds brought by Thonning in 1807 from Copenhagen Royal Botanic Garden.—F.N.H.

Solanum melongena *L.*—FTA. 4, 2: 242 (1906); FWTA. ed. 2, 2: 332 (1963).

S. edule Schum. et Thonn. in Schum. 1827: 125 and 1828: 145; Junghans 1962:91.

Exsicc.: Whydah, 1785, *Isert* s.n. (**C**—1 sheet, microf. 102: II. 4); "Guinea", *Thonning* 141 (**C**—1 sheet, microf. 102: II. 5, 6).

Vernacular name: 'Blafo-atropo'.

"Is cultivated."
"It is closely related to *S. melongena* and *S. insanum* but nevertheless different from both."
"Is used in same way as the preceding [*S. macrocarpon*] both by Europeans and natives."

Solanum nigrum *L.*—FWTA. ed. 2, 2: 335 (1963).

S. nodiflorum Jacq.—Schum. 1827: 123 and 1828: 143; Junghans 1962:91.

Exsicc.: "Guinea", *Thonning* 134 (**C**—4 sheets, microf. 102: III. 1–5; **P–JU**—1 sheet, No. 6376A, microf. 460: III. 3).

Vernacular name: 'Dendrae'.

"Here and there."
"The leaves are used against rheumatic pains; they are pounded with some grains of Paradise [*Aframomum*] into an ointment or salve, which is rubbed on the painful places."

Note: *S. nigrum* is here used in the traditional very broad sense. Recent work by Jennifer M. Edmonds (in J. Arn. Arb. 52: 634 (1971) and in Kew Bull. 27: 95–114 (1972) is elucidating the complex taxonomy.—F.N.H.

SPHENOCLEACEAE

Sphenoclea zeylanica *Gaertn.*—Schum. 1827: 103 and 1828: 123 (no MS.); FWTA. ed. 2, 2: 307, fig. 272; Junghans 1961: 327.

EXSICC.: Whydah, 1785, *Isert* s.n. (**C**—1 sheet, type, microf. 104: I. 5).

"Grows at Fida".

STERCULIACEAE

Cola gigantea *A. Chev.* var. **glabrescens** *Brenan et Keay*—FWTA. ed. 2, 1: 330 (1958).

Moraceae indet. in Junghans 1961: 345.

EXSICC.: Whydah, 1784, *Isert* s.n. (**C**—1 sheet, microf. 69: III. 2).

Cola verticillata (*Thonning*) *Stapf ex A. Chev.*—FWTA. ed. 2, 1: 330 (1958).

Sterculia verticillata Thonning in Schum. 1827: 240 and 1829: 14; Junghans 1962: 92.

EXSICC.: Aquapim, *Thonning* 283 (**C**—2 sheets, types, microf. 105: I. 3–6); "e Guinea" (**P–JU**—1 sheet, No. 12443, microf. 921: II. 7).

VERNACULAR NAME: 'Kjaelae'.

"Here and there at Aquapim."
"The fruit is chewed by the aborigines; it has a bitter astringent taste and colours the saliva carmine-red. Th."

Melochia corchorifolia *L.*—Schum. 1827: 297 and 1829: 71 (no MS.); FWTA. ed. 2, 1: 318 (1958); Junghans 1962: 92.

EXSICC.: Whydah, 1785, *Isert* s.n. (**C**—1 sheet, microf. 67: I. 5); "Guinea", *Thonning* s.n. (**C**—1 sheet, microf. nil; **P–JU**—1 sheet, No. 12490, microf. 924: II. 3).

"Here and there."
"The Indian plant has quite smooth and smaller leaves; but otherwise agrees with our description. S."

Melochia melissifolia *Benth.* var. **bracteosa** (*F. Hoffm.*) *K. Schum.*—FWTA. ed. 2, 1: 318 (1958).

Sterculiaceae indet. in Junghans 1962: 92 (see also *Chasmanthera dependens*).

EXSICC.: Whydah, 1785, *Isert* s.n. (**C**—1 sheet, microf. 67: I. 6).

Waltheria indica *L.*—FWTA. ed. 2, 1: 319 (1958). *W. americana* L.—FTA. 1: 235 (1868).

W. africana Schumacher 1827: 296 and 1829: 70; Junghans 1962: 92.

Exsicc.: Ada, 1784, *Isert* s.n. (**C**—1 sheet, microf. 110: II. 7); "Guinea", *Thonning* 10 (**C**—2 sheets, types, microf. 110: III. 1–4).

"Here and there."

"Very different from *W. indica* which has ovate, shorter more obtuse leaves, the flower-balls much smaller and sessile. *W.elliptica* Cavanilles has elliptic leaves which are deeply serrated, obtuse, narrower at the base; the stem is moreover branched. S."

W. guineensis Schum. et Thonn. in Schum. 1827: 295 and 1829: 69; Junghans 1962: 92.

Exsicc.: "Guinea", *Thonning* 10 (**C**—3 sheets, types, microf. 110: III: 5, 6; 111: I. 1–4).

VERNACULAR NAME: 'Fufuba'.

"Here and there."

"Approaches *W. americana* but in the latter the leaves are more deeply and pointedly toothed, ovate, more pointed, less felted; the leaf-stalks shorter (scarcely half an inch) the flowering branches longer than the leaves divided in the tip, the flower-balls sessile or stalked, larger; the bracts of the flowers and the sepals longer. S."

"When the leaves are dried and used as tea they resemble in taste the flowers of Verbascum and have the same action. Th."

NOTE: Junghans 1962: 92 observes that these specimens have the same number (10), as *W. guineensis* "was separated from *W. africana* fairly late". However, they must both be regarded now as *W. indica*.—F.N.H.

TILIACEAE

Clappertonia ficifolia (*Willd.*) *Decne.*—FWTA. ed. 2, 1: 310 (1958).

Honckenya ficifolia Willd. 1793: 201, pl. 4 ("Habitat in Guinea"); 1799: 325; Junghans 1962: 92.

Exsicc.: Whydah, 1784, *Isert* s.n. (**B**—1 sheet, No. 7249, type, microf. 500: II. 2; **C**—2 sheets, types, microf. 54: II. 6, 7; III. 1).

Corchorus aestuans *L.*—FWTA. ed. 2, 1: 308 (1958).

C. polygonus Schum. et Thonn. in Schum. 1827: 245 and 1829: 19; Junghans 1962: 92.

Exsicc.: "Guinea", *Thonning* 139 (**C**—3 sheets, types, microf. 26: III. 5–7; 27: I. 1, only 2 sheets photographed); "e Guinea" (**P–JU**—1 sheet, No. 12524, microf. 926: II. 5).

VERNACULAR NAME: 'Koina-fye'.

"Here and there, flowers along with the foregoing [i.e. *C. tridens*]."

"Resembles *C. aestuans* but the latter is different in that the stem is rather rough, the capsules linear six-sided, styles 3 bipartite, teeth of capsule longer and more pointed. S."

"Used as the preceding [*C. tridens*]. Th."

Corchorus tridens *L.*—FTA. 1 : 264 (1868); FWTA. ed. 2, 1 : 308 (1958).

C. angustifolius Schum. et Thonn. in Schum. 1827: 244 and 1829: 18; Junghans 1962: 92.

EXSICC.: Ada, 1784, *Isert* s.n. (**C**—1 sheet, microf. 26: II. 2); "Guinea", *Thonning* 137 (**C**—5 sheets, types, microf. 26: I. 1–7; II. 1); "e Guinea" (**P–JU**—1 sheet, No. 12534, microf. 927: I. 5).

VERNACULAR NAME: 'Koina-fye'.

"Here and there. Flowers in the fruitful time of the year."
"The leaves are eaten by the natives as cabbage. Th."
"Related to *C. tridens* but the stipules in this latter are tripartite, the leaves undulately serrated, the capsules rough, the 3 styles spread out, bipartite. S."

Corchorus trilocularis *L.*—FWTA. ed. 2, 1 : 308 (1958).

C. muricatus Schum. et Thonn. in Schum. 1827: 246 and 1829: 20; Junghans 1962: 92.

EXSICC.: "Guinea", *Thonning* 138 (**C**—3 sheets, types, microf. 26: II. 3–7; III. 1); "e Guinea?" (**C**—1 sheet, microf. 26: III. 2–4; **FI**—1 sheet, No. 27790; **P–JU**—1 sheet, No. 12528, microf. 926: III. 5).

VERNACULAR NAME: 'Koina-fye'.

"Here and there, flowers in the fruitful period of the year."
"Used as the preceding ones. Th."
"Much resembles *C. acutangulus* but in the latter the capsules are prismatic-cuneate, with sharp angles, three-toothed, not obtuse. *C. siliquosus* differs with linear compressed 2-valved capsules and leaves without bristles. S."

Grewia carpinifolia *Juss.* 1804: 91, Pl. 51, f. 1 ("ex Guinea et Owaria"); P. Beauv. 1804: 50, pl. 30 (1805); Schum. 1827: 241 and 1829: 15; FWTA. ed. 2, 1 : 305 (1958); Junghans 1962: 92.

EXSICC.: "Guinea", 1786, *Isert* s.n. (**C**—2 sheets, microf. 48: II. 6, 7; III. 2); "Guinea" coast, *Thonning* 24 (**C**—5 sheets, ?types, microf. 48: I. 7; II. 1–5; III. 1).

VERNACULAR NAME: 'Asi-gremi'.

"Frequent on sea coasts."
"The pulp of the berry is slightly acid; is eaten by children. The young shoots are used as cabbage. Th."

NOTE: Junghans recorded 4 sheets instead of the 5 now known—F.N.H.

Triumfetta rhomboidea *Jacq.*—FTA. 1 : 257 (1868); FWTA. ed. 2, 1 : 309 (1958).

T. thonningiana DC. 1825–26: 64.

EXSICC.: "Guinea", *Thonning* s.n. (**G–DC**—1 sheet, No. 1 : 507. 19, microf. nil).

T. mollis Schum. et Thonn. in Schum. 1827: 238 and 1829: 12; Junghans 1962: 93.

Exsicc.: "Guinea", *Thonning* 25 (**C**—7 sheets, types, microf. 107: I. 4–7; II. 1–7; III. 1); "e Guinea" (**C**—1 sheet, microf. 107: III. 2, 3; **P–JU**—1 sheet, No. 12542, microf. 927: III. 6).

Vernacular name: 'Toubé'.

"Common".
"Resembles *T. semitriloba* but in this latter the branches are less felted, the leaves downy-haired not felted; lobes and teeth sharper, the flowers larger. Perhaps *T. thonningiana* Decandolle is the same plant see Ferussae Bulletin, 1826. Sept. p. 52. S."
"The root is used for sores from Guinea worms. Th."

TURNERACEAE

Wormskioldia pilosa (*Willd.*) *Schweinf. ex Urb.*—FWTA. ed. 2, 1: 85 (1954).

Raphanus pilosus Willd. 1800b: 562 ("Habitat in Guinea"); Junghans 1962: 93.

Exsicc.: "Guinea", *Isert* s.n. (no type specimen traced at **B**).

Cleome raphanoides DC. 1824: 240. *Tricliceras raphanoides* (DC.) DC. 1825–26: 56.

Exsicc.: "Guinea", *Isert* s.n. (**G–DC**—1 sheet, 1: 240. 24, type, microf. 121: II. 6); "Guinea", *Thonning* 297 (**G–DC**—1 sheet, 1: 240. 24, type, microf. 121: II. 7).

Wormskioldia heterophylla Schum. et Thonn. in Schum. 1827: 165 and 1828: 185; FTA. 2: 502 (1871).

Exsicc.: "Guinea", *Thonning* 297 (**C**—1 sheet, type, microf. 111: II. 1, 2).

"Fairly rare in dry fields; flowers in May."
"I have named this genus after Morten Wormskiold, Knight of Danne-brog, known for his travels in Greenland, Kamtschatka and in the South Seas. S."

ULMACEAE

Trema orientalis (*L.*) *Blume*—FTEA. Ulmac. 10 (1966).

Celtis guineensis Schum. et Thonn. in Schum. 1827: 160 and 1828: 180; Junghans 1962: 93. *T. guineensis* (Schum. et Thonn.) Ficalho—FTA. 6, 2: 11 (1916); FWTA. ed. 2, 1: 592 (1958).

Exsicc.: "Guinea", *Thonning* 218 (**C**—7 sheets, types, microf. 17: II. 4–7; III. 1–6).

"Here and there among other shrubs."
"Our plant approaches *C. orientalis* L. and *C. crassifolia* Lamarck; but the

first named (according to a specimen in Hermann's collection) have somewhat hairy branches, slightly narrower and much more acuminate leaves
with smaller serrations; the leaves are further grey below, almost silky-
haired; the panicles have more divergent branches and are slightly longer
than the leaf-stalks. *C. crassifolia* differs in having the leaves cordate-oblong
at the base, oblique and undulate, whitish more hairy leaf-stalks, and
smooth branches. S."

C. guineensis var. *β parvifolia* Schumacher 1827: 161 and 1828: 181.

Exsicc.: "Guinea", *Thonning* 218 (**C**—1 sheet, type, microf. 17: III. 7; 18:
I. 1, 2)

"Agrees in the principal characters with the plant just described, but the
branches are more felted-haired, the leaves smaller, on the upper side rough
and dark green on the under side almost rusty coloured, the flowers are
more crowded. At first glimpse *Celtis lima* rather closely resembles this
variety but its leaves are oblong-lanceolate, more still, rusty coloured and
felted; the flower balls in the leaf axils are many-flowered. S."

UMBELLIFERAE

Centella asiatica (*L.*) *Urb.*—FWTA. ed. 2, 1: 753 (1958).

Exsicc.: Whydah, 1785, *Isert* s.n. (**C**—1 sheet, microf. nil).

Note: This sheet has been found since Junghans prepared his paper.—
F.N.H.

URTICACEAE

Laportea ovalifolia (*Schum. et Thonn.*) *Chew* in Gard. Bull. Singapore 21:
201 (1965). *Fleurya ovalifolia* (Schum. et Thonn.) Dandy—FWTA. ed. 2, 1:
619 (1958). *F. podocarpa* Wedd.—FTA. 6, 2: 251 (1917).

Haynea ovalifolia Schum. et Thonn. in Schum. 1827: 406 and 1829: 180;
Keay in Kew Bull. 12: 175–176 (1957); Junghans 1962: 93.

Exsicc.: Aquapim, *Thonning* 264 (see note after *Urera mannii*) (**C**—1 sheet,
type, microf. 49: II. 3, 4).

"Near Aquapim."
"As Wildenow's *Haynea* is named by Persoon and others with Aublet's
older name, I have named it anew after Professor Hayne in Berlin. S."

Urera mannii (*Wedd.*) *Benth. et Hook. f. ex Rendle*—FWTA. ed. 2, 1: 618
(1958).

Urticaceae indet. in Junghans 1962: 93.

Exsicc.: Aquapim?, *Thonning* 264? (**C**—3 sheets, microf. 49: I. 4–7; II. 1,
2).

Note: These sheets are written up as "Thonning 264 Haynea ovalifolia" but

R. W. J. Keay noted on the sheets in 1955 that they differ seriously from the description of that species and cannot be taken as the type.—F.N.H.

VAHLIACEAE see SAXIFRAGACEAE

VERBENACEAE

Avicennia germinans (*L.*) *L.*—Compère in Taxon 12: 150 (1963).

A. africana P. Beauv.—Schum. 1827: 290 and 1829: 64; FWTA. ed. 2, 2: 448 fig. 309 (1963); Junghans 1962: 94.

Exsicc.: Ada, 1784, *Isert* s.n. (**C**—2 sheets, microf. 10: I. 5, 7); coast and R. Volta, *Thonning* 98? (**C**—1 sheet, microf. 10: I. 6).

Vernacular name: 'Muteku'.

"Grows near the small salt areas which occur near the coast, but rarely over 6–8 ells in height; near Rio Volta on the other hand and in areas near by, where the sea wind has not such a strong influence, it is as large as the biggest oak-trees. Flowers commonly at the beginning of the rainy season; but also at other times of the year."

"The trunk consists of two parts; the internal or so called heart of the tree has a dark brownish green colour almost like the guayac tree; it is hard, heavy, coarse and contains much resin, and has no distinct odour or taste. The resin is by no means abundant as to exude; this internal portion constitutes slightly over half the diameter of the tree; this is very firm for coarse work and mainly in the ground. Commander Schönning assured me that near Kongesteen, he had had a stake dug up which was almost stone-hard; perhaps the salty nature of the ground near Volta considerably contributed to this. The tree cannot be rough-hewed with an axe without care as it easily splits awry. The external portion of the trunk has the usual light yellowish tree-colour and is much less durable. The natives as far as I know make no medicinal use of this tree. Th."

"*Avicennia tomentosa* (from India) differs from our plant in having ovate or ovate-oblong leaves, which are 2 or scarcely 3 inches long. The twigs and flower-stalks are finer, the flowers and leaves of the same less so. *A. tomentosa* from Arabia (*Sceura maritima* Forsk.) differs in having elliptic-lanceolate pointed leaves. The American plant similarly, and these two last should perhaps be separated from the Indian one. S."

Clerodendrum capitatum (*Willd.*) *Schum. et Thonn.* var. **capitatum**— Schum. 1827: 287 and 1829: 61; FWTA. ed. 2, 2: 443 (1963); Junghans 1962: 94.

Volkameria capitata Willd. 1800b: 384 ("Habitat in Guinea"); Junghans 1962: 94.

Exsicc.: Aquapim, 1786, *Isert* s.n. (**B**—1 sheet, No. 11682, type, microf. 831: II. 7; **C**—1 sheet, type, microf. 21: II. 2); Dudun, *Thonning* 225 (**C**—1 sheet, microf. 21: I. 7; II. 1; **P–JU**—1 sheet, No. 5027, microf. 347: II. 5).

"In Aquapim and in Dudun."

Lantana camara *L.*—FWTA. ed. 2, 2: 435 (1963).

L. antidotalis Schum. et Thonn. in Schum. 1827: 276 and 1829: 50; FTA. 5: 276 (1900); Junghans 1962: 94.

EXSICC.: "Guinea", *Thonning* 125 (**C**—2 sheets, types, microf. 64: I. 3–7; II. 1; **P–JU**—1 sheet, No. 5116, type, microf. 354: II. 2).

VERNACULAR NAME: 'Nanni-kumi'.

"Grows here and there. Flowers nearly the whole year through but especially in the rainy season."

"The leaves have a powerful odour. When a native is bitten by a snake which usually occurs in the foot, he immediately tries to stop the circulation of blood by tying a band round the leg as tight as possible; he then looks for a doctor who orders certain superstitious treatments and the use of some medicinal plants. In this case the leaves of this plant are often used in a warm bath, and the root is finely macerated with grains-of-Paradise [*Aframomum*] and lemon-juice and rubbed over the whole body, so that the poison shall not spread. At the same time the doctor has some medicine prepared for internal use which I do not know. The root of this shrub will be intermixed with it. Finally he performs some conjuring tricks in order in a secret way to get rid of the snake's teeth, which the natives believe are in the wound. Th."

"It much resembles *L. aculeata* but in the latter the leaves are more cordate, rough and wrinkled; the serrations are smaller, the prickles of the tsem larger, more frequent and recurved. S."

Premna quadrifolia *Schum. et Thonn.* in Schum. 1827: 275 and 1829: 49; FWTA. ed. 2, 2: 438 (1963); Junghans 1962: 94.

EXSICC.: Frederiksberg, *Thonning* 161 (**C**—3 sheets, types, microf. 89: I. 7; II. 1–5); "e Guinea" (**G–DC**—1 sheet, 11: 633. 13, microf. 1899: II. 6; **P–JU**—1 sheet, No. 5075, microf. 351: I. 5).

VERNACULAR NAME: 'Oobsso-tjo'.

"Found once near Frederiksberg. Flowers in January."

"The leaves and even more, the fruit have a strong unpleasant odour. The leaves pounded in tepid water are used as a laxative. A bottle-shaped calebash is used for this purpose, whose tube is inserted, after which the doctor blows with his mouth through an opening in the bottom. Th."

"*P. serratifolia* differs from our plant in having opposite, ovate, serrated, smooth leaves. S."

Stachytarpheta angustifolia (*Mill.*) *Vahl*—FWTA. ed. 2, 2: 434 (1963).

S. indica sensu Schum. 1827: 14 and 1828: 34; Junghans 1962: 94.

EXSICC.: "Guinea", *Isert* s.n. (**C**—3 sheets, microf. 104: I. 6, 7; II. 1, 2, 5, 6); "Guinea", *Thonning* 306 (**C**—4 sheets, microf. 104: II. 3, 4, 7; III. 1; only 3 sheets photographed).

VERNACULAR NAME: 'Lalaaba'.

"Grows commonly on plains, where the soil is not dry and poor."

E

"The natives use this plant for inflammations of the eyes and spots on the cornea in the following way: the leaves are crushed between hot stones, and then placed in a linen cloth from which the sap is squeezed into the eye."

NOTE: All the material is of this species, although when Vahl wrote up his *S. indica* (Mill.) Vahl 1804: 206 he gave the distribution as "Habitat in Zeylona et Guinea". Schumacher apparently confused the species.—F.N.H.

Vitex doniana *Sweet*—FWTA. ed. 2, 2: 446 (1963).

V. cuneata Thonning in Schum. 1827: 289 and 1829: 63; FTA. 5: 238 (1900); Junghans 1962: 94.

EXSICC.: Sawi, 1785, *Isert* s.n. (**C**—1 sheet, microf. 109: II. 7); "Guinea", *Thonning* 244 (**C**—2 sheets, types, microf. 109: III. 1–4; **LE**—1 sheet, type).

VERNACULAR NAME: 'Fjong'.

"Common."

"The fruit is eaten by the inhabitants although for those unaccustomed to it, it has not a pleasant taste; the pulp is half sappy mealy, slightly sweet with an oily taste. The wood is used by the natives for drums. Th."

"In the shape of the leaves it is different enough from *V. leucoxylon*. S."

Verbenaceae indet. in Junghans 1962: 94.

EXSICC.: Guinea, *Isert* s.n. (**C**—1 sheet, microf. 109: III. 5).

Vitex ferruginea *Schum. et Thonn.* in Schum. 1827: 288 and 1829: 62; FWTA. ed. 2, 2: 477 (1963); Junghans 1962: 94.

EXSICC.: Aquapim, *Thonning* 265 (**C**—3 sheets, types, microf. 109: III. 6, 7; 110: I. 1–4); "e Guinea" (**FI**—1 sheet; **P–JU**—1 sheet, No. 5052, type, microf. 349: I. 6).

"Near Aquapim."

"*V. leucoxylon* differs from our plant in having elliptical-oblong rather obtuse, quite smooth long-stalked leaves and dichotomously-divided long-stalked corymbs. S."

VIOLACEAE

Hybanthus enneaspermus *(L.) F. v. Müll.*—FWTA. ed. 2, 1: 106 (1954). *Ionidium ennaespermum* Vent.—FTA. 1: 105 (1868), partly.

Viola lanceifolia Thonning in Schum. 1827: 134 and 1828: 154; Junghans 1962: 95.

EXSICC.: Dudua and Quita, *Thonning* 219 (**C**—5 sheets, types, microf. nil).

"Rare, found at Dudua and Quita."

Hybanthus thesiifolius *(Juss. ex Poir.) Hutch. et Dalz.*—FWTA. ed. 2, 1: 106. *Ionidium ennaespermum* Vent.—FTA. 1: 105 (1868), partly.

Viola guineensis Schum. et Thonn. in Schum. 1827: 133 and 1828: 153; Junghans 1962: 95.

Exsicc.: "Guinea", *Isert* s.n. (**C**—1 sheet, microf. nil); "Guinea", *Thonning* 73 (**C**—2 sheets, types, microf. nil).

"Common in open fields."
"In the characters cited it is easily distinguished from *V. linearifolia* Vahl. Eclog. fas. 2, p. 18 and from *V. suffruticosa* and *V. enneasperma*. S."

VITACEAE

Cyphostemma cymosa (*Schum. et Thonn.*) *Descoings* in Notul. Syst. Paris 16: 121 (1960). *Vitis thonningii* Baker in FTA. 1: 407 (1868)—based on this species with the same type.

Cissus cymosa Schum. et Thonn. in Schum. 1827: 82 and 1828: 102; FWTA. ed. 2, 1: 680 (1958); Junghans 1961: 322.

Exsicc.: Ada 1784, *Isert* s.n. (**C**—1 sheet, microf. 20. II. 4); "Guinea", *Thonning* 179 (**C**—2 sheets, types, microf. 20: II. 5–6); "e Guinea" (**P-JU**—1 sheet, No. 11997, microf. 888: III. 3).

Vernacular name: 'Anmanum-Ba'.

"Here and there in thicket. Flowers in June and July."

Cissus quadrangularis *L.*—FTA. 1: 399 (1868); FWTA. ed. 2, 1: 676 (1958).

Cissus bifida Schum. et Thonn. in Schum. 1827: 80 and 1828: 100; Junghans 1961: 322.

Exsicc.: "Guinea", *Thonning* 301 (**C**—1 sheet, microf. 20: II. 3).

"Here and there."
"In habit it resembles *C. quadrangularis* or *Sael nthus quadragonus* Forsk. Descrip. p. 33 & icon. 2, but differs in that the setae situated on the corners of the stem-segment are cleft, leaves, leaf-stalks and flower-stalks hairy. The leaves crushed and mixed with lemon juice are applied to swellings caused by Guinea worms, Filaria medinensis. This medicine is also used for treating the illness the inhabitants call Anasarea (a kind of aquaeous sickness), when it is mixed with crushed grains-of-Paradise [*Aframomum*] and rubbed into the whole body. Th."

Cissus triandra Schum. et Thonn. in Schum. 1827: 81 and 1828: 101; Junghans 1961: 322.

Exsicc.: Gah and Adampi, *Thonning* 292 (**C**—2 sheets, types, microf. 20: II. 7; III. 1).

"Grows, although rarely in thicket, in the Gah and Adampi areas."

Leea guineensis *G. Don*—FWTA. ed. 2, 1: 683 (1958).

L. sambucina sensu Schum. 1827: 134 and 1828: 154; FTA. 1: 415 (1868); Junghans 1961: 322, non (L.) Willd.

Exsicc.: Aquapim, *Thonning* 280 (**C**—1 sheet, microf. 64: II. 2–3).

E 2

"Here and there in Aquapim, preferably in moist places."

"The Guinea plant differs from the Indian one merely in having broader and larger leaflets."

NOTE: The single sheet is labelled simply "e Guinea", but as it is the only one in the herbarium there is no doubt it is Thonning's specimen No. 280.— F.N.H.

ZYGOPHYLLACEAE

Tribulus terrestris *L.*—FTA. 1 : 284 (1868); FWTA. ed. 2, 1 : 363 (1958).

T. humifusus Schum. et Thonn. in Schum. 1827: 215 and 1828: 235; Junghans 1962: 95.

EXSICC.: "Guinea", *Thonning* 30 (**C**—1 sheet, type, microf. 107: I. 2, 3).

VERNACULAR NAME: 'Bláfo' (Accra).

"Grows here and there."

"In the characters cited it adequately differs from *T. terrestris*, *T. lanuginosus* and *T. cystoides*. S."

"A dangerous weed on footpaths. Th."

MONOCOTYLEDONS

AGAVACEAE

Sansevieria liberica *Gér. et Labr.*—FWTA. ed. 2, 3: 159 (1968).

S. guineensis sensu Willd. 1799: 159; Schum. 1827: 174 and 1828: 194; Junghans 1962: 95.

EXSICC.: "Guinea", *Thonning* 238 (**C**—1 sheet, microf. 94: III. 4–6).

VERNACULAR NAME: 'Blaa'.

"Here and there in coastal areas."

"Fishermen collect the leaves in quantity, and by soaking them in water and beating them, separate the fibrous and fleshy portions from each other; the former is a very good hemp, from which they prepare the coarsest cords or reefs for their fishing-nets. Adanson mentions the same use of this plant in Senegal. In the same way the natives prepare from Ananas (pineapple) leaves a fairly fine and long but somewhat coarse fibre, of which they prepare all their thread and yarn for sewing, fishing-nets, etc. They have even had another plant which yields hemp, which far surpasses the usual one in strength; the plant is subherbaceous, twining and jointed; the hemp consists actually in splinters, which are picked out with the fingers and are no longer than the joint or segment of the plant. Th."

NOTE: The specimen in Willdenow's Herbarium (No. 6711, microf. 466: II. 1.) is not labelled with Isert's name.—F.N.H.

ALISMATACEAE

Limnophyton obtusifolium (*L.*) *Miq.*—FTA. 8: 209 (1901); FTEA. Alismatac. 9 (1960); FWTA. ed. 2, 3: 11 (1968).

Alisma sagittifolia Willd. 1799: 277 ("Habitat in Guinea"); Schum. 1827: 182 and 1828: 202; Junghans 1962: 95.

Exsicc.: "Guinea", *Isert* s.n. (**B**—1 sheet, No. 7107, type, microf. 492: I. 4— see note); R. Laloe, *Thonning* 152 (no material traced).

"At the river Laloe; rare."

Note: The specimen in Willdenow's Herbarium appears on an incorrectly numbered microfiche: 492 should read 491. There are also 2 other specimens numbered 7107 which should both read 7106a.—F.N.H.

AMARYLLIDACEAE

Amaryllis nivea *Thonning* in Schum. 1827: 169 and 1828: 189; Junghans 1962: 95.

Exsicc.: Aquapim, *Thonning* 169 (no type specimen traced).

"In Aquapim's woods."
"It differs from *A. zeylanica* in that the leaf margin is smooth and not dentately rough; the corollas white without a red keel. Th."

Note.: This is almost certainly *Crinum jagus* (Thomps.) Dandy (1939), see below, based on *Amaryllis jagus* Thomps. (1798).—F.N.H.

Amaryllis trigona *Thonning* in Schum. 1827: 170 and 1828: 190; Junghans 1962: 95.

Exsicc.: "Guinea", *Thonning* 39 (no type specimen traced).

"Common in coastal areas; flowers in May and June."
"Related to *A. ornata* and perhaps not different from it. S."
"The bulb acts as an epispasticum. Th."

Note.: This is almost certainly *Crinum ornatum* (Ait.) Bury (1834)—FWTA. ed. 2, 3: 134 (1968), based on *Amaryllis ornata* Ait. (1789).—F.N.H.

Crinum jagus (*Thomps.*) *Dandy*—FWTA. ed. 2, 3: 136 (1968).

Amaryllidaceae indet. in Junghans 1962: 96.

Exsicc.: Sawi, 1785, *Isert* s.n. (**C**—1 sheet, microf. 5: I. 4).

Haemanthus multiflorus *Martyn*—FWTA, ed. 2, 3: 132 (1968).

Amaryllidaceae indet. in Junghans 1962: 96.

H. cruentatus Schum. et Thonn. in Schum. 1827: 168 and 1828: 188; Junghans 1962: 96.

Exsicc.: Ada, 1784, *Isert* s.n. (**C**—1 sheet, holotype det. I. Björnstad, microf. 49: I. 3); "Guinea", *Thonning* 257 (no type specimen at **C**; **P–JU**—1 sheet, No. 3479, microf. 229: III. 3).

Vernacular name: 'Mihaa' or 'Maej'.

"Here and there in coastal areas. Flowers in April and produces leaves in June."
"Resembles *H. multiflorus* but differs in the cited characters."

Pancratium tenuifolium *Hochst. ex A. Rich.*—Geerinck in Fl. Afr. Centr., Amaryl. 20 (1973). *P. hirtum* A. Chev.—FWTA. ed. 2, 3: 136 (1968).

Amaryllidaceae indet. in Junghans 1962: 96.

Exsicc.: Sawi, 1787, *Isert* s.n. (**C**—1 sheet, microf. 5: I. 3).

APONOGETONACEAE

Aponogeton subconjugatus *Schum. et Thonn.* in Schum. 1827: 183 and 1828: 203; FTA. 8: 217 (1901); FWTA. ed. 2, 3: 15 (1968); Junghans 1962: 96.

Exsicc.: "Guinea", *Thonning* 103 (**C**—1 sheet, type, microf. 6: II. 2); "e Guinea" (**FI**—1 sheet; **P–JU**—1 sheet, No. 17374, microf. 1255: II. 4).

"In stagnant water; flowers in June."
"It is doubtful if it is different from *A. monostachyos*. S."

ARACEAE

Colocasia esculenta *(L.) Schott*—FWTA. ed. 2, 3: 119 (1968).

Caladium esculentum (L.) Vent.—Schum. 1827: 408 and 1829: 182; Junghans 1962: 96.

Exsicc.: (No material traced).

"Cultivated."

Cyrtosperma senegalense *(Schott) Engl.*—FWTA. ed. 2, 3: 113 (1968).

Araceae indet. in Junghans 1962: 96.

Exsicc.: Whydah, 1785, *Isert* s.n. (**C**—1 sheet, microf. 34: III. 6).

Pistia stratiotes *L.*—Schum. 1827: 298 and 1829: 72; FWTA. ed. 2, 3: 113 (1968); Junghans 1962: 96.

Exsicc.: Fida, 1784, *Isert* s.n. (**C**—1 sheet, microf. 85: I. 7; II. 1).

CANNACEAE

Canna indica *L.*—FWTA. ed. 2, 3: 79 (1968).

C. rubra Willd.—Schum. 1827: 395 and 1829: 169; Junghans 1962: 96.

Exsicc.: Aquapim, *Thonning* s.n. (**C**—1 sheet, microf. 13: III. 4, 5).

"Here and there in Aquapim."

COMMELINACEAE

Aneilema umbrosum *(Vahl) Kunth*—FWTA. ed. 2, 3: 30 (1968). *A. ovato-oblongum* P. Beauv.—FTA. 8: 69 (1901).

Commelina umbrosa Vahl 1805: 179 ("Habitat in umbrosis Assjamac Guinea. *Isert*"); Schum. 1827: 23 and 1828: 43; Junghans 1962: 96.

Exsicc.: Asohia (? Assiamae), 1786, *Isert* s.n. (**C**—2 sheets, types, microf. 23: I. 2–4); Aquapim, *Thonning* 206 (**C**—1 sheet, microf. 23: I. 1).

"Grows in Aquapim according to Thonning, in Assiamae in very shady places according to Isert."
"Although Vahl's and Thonning's descriptions do not correspond, I am quite sure that both of them were made from the same species but Thonning made his at the growing place using the fresh plant whereas Vahl used dry specimens." [S].

Commelina erecta *L.*—FWTA. ed. 2, 3: 49 (1968).

C. umbellata Thonning in Schum. 1827: 21 and 1828: 41; FTA. 8: 55 (1901); Junghans 1962: 96.

Exsicc.: "Guinea", *Thonning* 74 (**C**—4 sheets, type, microf. 22: III. 2–7); "e Guinea" (**P–JU**—1 sheet, No. 3209, microf. 211: II. 5).

"Common".

Murdannia simplex *(Vahl) Brenan*—FWTA. ed. 2, 3: 24 (1968). *Commelina sinicum* (Roem. et Schult.) Lindl. var. *simplex* (Vahl) C.B. Cl.—FTA. 8: 64 (1901).

Commelina simplex Vahl 1805: 177 ("Habitat in Guinea. *Isert, Thonning*); Schum. 1827: 22 and 1828: 42, 43; Junghans 1962: 96.

Exsicc.: "Guinea"; *Isert* s.n. (**C**—1 sheet, type, microf. 22. III. 1); "Guinea", *Thonning* 75 (**C**—4 sheets, type, microf. 22: II. 1–7).

"Grows in dry sandy and open fields."

CYPERACEAE*

Ascolepis dipsacoides *(Schumacher) J. Raynal*—FWTA. ed. 2, 3: 326 (1972).

Kyllinga dipsacoides Schumacher 1827: 41 and 1828: 61; Junghans 1962: 99.

Exsicc.: "Guinea", *Thonning* 404 (**C**—2 sheets, types, microf. 62: III. 3–6); "e Guinea" (**C**—1 sheet, microf. 63: I. 1, 2).

* Determinations revised by Miss S. S. Hooper (Kew).

Bulbostylis barbata (*Rottb.*) *C.B. Cl.*—FTA. 8: 431 (1901); FWTA. ed. 2, 3: 316 (1972).

Scirpus antarcticus sensu Vahl 1805: 261, non L.; Schum. 1827: 29 and 1828: 49, 50; Junghans 1962: 100.

Exsicc.: "Guinea", 1786, *Isert* s.n. (**C**—1 sheet, microf. 96: II. 4, 5); "Guinea", *Thonning* 378 (**C**—3 sheets, microf. 96: II. 6, 7; III. 1–4); "e Guinea" (**P–JU**—1 sheet, No. 1780, microf. 116: II. 1).

Bulbostylis filamentosa (*Vahl*) *C.B. Cl.*—FTA. 8: 433 (1902); FWTA. ed. 2, 3: 317 (1972).

Scirpus filamentosa Vahl 1805: 262 ("Habitat in Guinea. *Thonning*"); Schum. 1827: 30 and 1828: 50; Junghans 1962: 100.

Exsicc.: "Guinea", *Thonning* s.n. (**C**—3 sheets, types microf. 97: I. 1–5); "e Guinea" (**C**—2 sheets, microf. 96: III. 5–7; **P–JU**—1 sheet, No. 1782 bis, microf. 122: I. 6).

Bulbostylis pilosa (*Willd.*) *Cherm.*—FWTA. ed. 2, 3: 316 (1972). *Fimbristylis africana* Dur. et Schinz.—FTA. 8: 425 (1902).

Schoenus pilosus Willd. 1794: 3, pl. 1, f. 3 ("Habitat in Guinea. *Isert*"); Vahl 1805: 208; Schum. 1827: 29 and 1828: 49; Junghans 1962: 100.

Exsicc.: "Guinea", 1784–5, *Isert* s.n. (**B**—1 sheet, No. 1095, type, microf. 63: I. 2; **C**—2 sheets, types, microf. 95: II. 5–7; III. 1); "Guinea", *Thonning* 393 (**C**—4 sheets, microf. 95: I. 4–7; II. 1–4); "e Guinea" (**C**—1 sheet, microf. 95: I. 3; **P–JU**—1 sheet, No. 1721, microf. 112: III. 2).

Cyperus amabilis *Vahl* 1805: 318 ("Habitat in Guinea. *Thonning*"); Schum. 1827: 35 and 1828: 55; FWTA. ed. 2, 3: 291 (1972); Junghans 1962: 97.

Exsicc.: "Guinea", *Thonning* 403 (**C**—3 sheets, types, microf. 30: II. 7; III. 1–5); "e Guinea" (**C**—2 sheets, microf. 30: III. 6–7; **FI**—1 sheet; **P–JU**—1 sheet, No. 1870 bis, microf. 122: I. 4).

C. microstachyos Vahl 1805: 318 ("Habitat in Guinea. *Thonning*"); Schum. 1827: 36 and 1828: 56 (no MS No.); FTA. 8: 328 (1901); Junghans 1962: 98.

Exsicc.: "Guinea", *Thonning* s.n. (or 350 according to Kükenthal) (**C**—1 sheet, type, microf. 33: I. 3, 4); "e Guinea" (**C**—2 sheets, microf. 33: I. 1, 2, 5, 6).

Cyperus aristatus *Rottb.*—Schum. 1827: 39 and 1828: 59; FWTA. ed. 2, 3: 294 (1972); Junghans 1962: 97.

Exsicc.: "Guinea", *Thonning* 405 (**C**—2 sheets, microf. 31: I. 5–7; II. 1).

"From this description (by Vahl, cited in the book) our plant differs in that the leaves are shorter, two of the bracts longer than [the peduncles] the other two of the same length as the peduncles, the umbel four-fid, the length of two of the peduncles three inches."

Cyperus articulatus *L.*—Vahl 1805: 301; FWTA. ed. 2, 3: 285 (1972); Junghans 1962: 97.

Exsicc.: Atacco, 1784, *Isert* s.n. (**C**—1 sheet, microf. 31: II. 2, 3).

Cyperus bulbosus *Vahl* 1805: 342 ("Habitat ad Senegal, in Guinea in India orientale. *Jussieu, Thonning, König*"); FWTA. ed. 2, 3: 286 (1972).

Exsicc.: see above (no type material at **C** or **P–JU**).

C. polyphyllus Vahl 1805: 317 ("Habitat in Guinea. *Thonning*"); Schum. 1827: 35 and 1825: 55; Junghans 1962: 98.

Exsicc.: "Guinea", *Thonning* 388 (**C**—2 sheets, types, microf. 33: III. 1–4); "e Guinea" (**C**—2 sheets, microf. 33: III. 5–7).

Cyperus difformis *L.*—Vahl 1805: 337; FWTA. ed. 2, 3: 290 (1972); Junghans 1962: 97.

Exsicc.: Whydah, 1785, *Isert* s.n. (**C**—1 sheet, microf. 31: III. 3, 4); "e Guinea" (**C**—1 sheet, microf. 31: III. 1, 2).

Cyperus dilatatus *Schumacher* 1827: 38 and 1828: 58; FWTA. ed. 2, 3: 286 (1972); Junghans 1962: 97.

Exsicc.: "Guinea", *Thonning* 397 (**C**—2 sheets, types, microf. 31: III. 5, 6, 7); "e Guinea" (**C**—1 sheet, microf. 32: I. 1, 2; **P–JU**—1 sheet, No. 1870 ter, microf. 122: I. 3).

"Related to *C. rotundus* but sufficiently different from it in the characters referred to."

Cyperus distans *L. f.*—Vahl 1805: 362; FWTA. ed. 2, 3: 287 (1972); Junghans 1962: 98.

Exsicc.: "Guinea", *Thonning* s.n. (**C**—1 sheet, microf. 32: I. 3, 4).

Cyperus fenzelianus *Steud.*—FWTA. ed. 2, 3: 285 (1972).

Cyperaceae indet. in Junghans 1962: 101.

Exsicc.: "Guinea", *Thonning* s.n. (**C**—1 sheet, microf. 34: III. 5).

Cyperus imbricatus *Retz.*—FWTA. ed. 2, 3: 284 (1972).

C. radiatus Vahl 1805: 369 ("Habitat in India orientali *König*, in Guinea *Isert*"); FTA. 8: 369 (1901); Junghans 1962: 99.

Exsicc.: Ada, 1784, *Isert* s.n. (**C**—1 sheet, type, microf. 34: I. 7; II. 1).

Note: Also the type sheet collected by König in India, appears on IDC No. 2203, microf. 34: I. 5, 6—F.N.H.

Cyperus margaritaceus *Vahl* 1805: 307 ("Habitat in Guinea. *Isert, Thonning*"); Schum. 1827: 33 and 1828: 53, 54; FTA. 8: 321 (1901);

FWTA. ed. 2, 3: 292 (1972); Junghans 1962: 98. *C. eburneus* Thonning ex Kunth 1837: 46.

Exsicc.: "Guinea", *Isert* s.n. (**B**—1 sheet, No. 1291, type, microf. 75: I. 4; **C**—1 sheet, type, microf. 32: III. 6, 7); "Guinea", *Thonning* 389 (**C**—4 sheets, types, microf. 32: II. 3, 4, 7; III. 1–5); "e Guinea" (**C**—1 sheet, microf. 32: III. 5, 6; **P–JU**—2 sheets, No. 1798, microf. 117: III. 6, 7).

Cyperus maritimus *Poir.*—FTA. 8: 327 (1901); FWTA. ed. 2, 3: 291 (1972).

C. scirpoides Vahl 1805: 311 ("Habitat in Guinea. *Thonning*"); Schum. 1827: 34 and 1828: 54; Junghans 1962: 99.

Exsicc.: "Guinea", *Thonning* 370 (**C**—2 sheets, types, microf. 34: II. 2–5).

Note: Clark and Kükenthal named this var. *crassipes* (Vahl) C.B. Cl.— F.N.H.

Cyperus nudicaulis *Poir.*—FTA. 8: 316 (1901); FWTA. ed. 2, 3: 293 (1972).

C. pectinatus Vahl 1805: 298 ("Habitat in Guinea. *Isert*"); Junghans 1962: 98.

Exsicc.: Whydah, 1785, *Isert* s.n. (**C**—1 sheet, type, microf. 33: II. 6, 7); "e Guinea?" (**C**—1 sheet, microf. 33: II. 4, 5).

Cyperus pustulatus *Vahl* 1805: 341 ("Habitat in Guinea. *Thonning*"); Schum. 1827: 37 and 1828: 57; FWTA. ed. 2, 3: 288 (1972); Junghans 1962: 98.

Exsicc.: "Guinea", *Thonning* 398 (**C**—2 sheets, types, microf. 34: I. 1–4).

Cyperus sphacelatus *Rottb.*—Vahl 1805: 341; Schum. 1827: 37 and 1828: 57; FWTA. ed. 2, 3: 286 (1972); Junghans 1962: 99.

Exsicc.: "Guinea", *Thonning* 399 (**C**—2 sheets, microf. 34: II. 6, 7; III. 1, 2); "e Guinea" "No. 228" (**C**—1 sheet, microf. 34: III. 3, 4).

Eleocharis acutangula *(Roxb.) Schult.*—FWTA. ed. 2, 3: 314 (1972).

Cyperaceae indet. in Junghans 1962: 101.

Exsicc.: Whydah, 1785, *Isert* s.n. (**C**—1 sheet, microf. 39: II. 4).

Fimbristylis complanata *Link*—FWTA. ed. 2, 3: 323 (1972).

Cyperaceae indet. in Junghans 1962: 101.

Exsicc.: Whydah, 1787, *Isert* s.n. (**C**—1 sheet, microf. 94: II. 3, 4).

Fimbristylis hispidula *(Vahl) Kunth*—FWTA. ed. 2, 3: 324 (1972).

Scirpus hispidulus Vahl 1805: 276 ("Habitat in Guinea. *Thonning*"); Schum. 1827: 31 and 1828: 51; FTA. 8: 418 (1902); Junghans 1962: 101.

Exsicc.: "Guinea", *Thonning* 349 (**C**—1 sheet, type, microf. 97: I. 7; II. 1); "e Guinea" (**C**—1 sheet, microf. 97: I. 6; **P–JU**—1 sheet, No. 1779, microf. 116: I. 7).

Fimbristylis obtusifolia *(Lam.) Kunth*—FWTA. ed. 2, 3: 324 (1972).

Scirpus obtusifolius Lam.—Vahl 1805: 275; Schum. 1827: 30 and 1828: 50; FTA. 8: 423 (1902); Junghans 1962: 101.

Exsicc.: "Guinea", *Thonning* 376 (**C**—2 sheets, microf. 97: II. 2–5).

Fimbristylis pilosa *Vahl* 1805: 290 ("Habitat in Insula Franciae, in Guinea. *Thonning*"); Schum. 1827: 32 and 1828: 52; FTA. 8: 416 (1902); FWTA. ed. 2, 3: 321 (1972); Junghans 1962: 99.

Exsicc.: "Guinea", *Isert* s.n. (**C**—1 sheet, microf. 42: III. 3, 4); "Guinea", *Thonning* 391 (**C**—2 sheets, types, microf. 42: II. 6, 7; III. 1, 2); "e Guinea" (**P–JU**—1 sheet, No. 1782, microf. 116: II. 3).

"The specimen which occurs in Vahl's herbarium from Ile de France under this name, is certainly another plant."

Fimbristylis scabrida *Schum.* 1827: 32 and 1828: 52; FWTA. ed. 2, 3: 323 (1972); Junghans 1962: 99.

Exsicc.: "Guinea", *Thonning* 394 (**C**—1 sheet, type, microf. 42: III. 5, 6).

Fimbristylis triflora *(L.) K. Schum.*—FWTA. ed. 2, 3: 324 (1972). *F. tristachya* Thw.—FTA. 8: 424 (1902).

Abildgaardia lanceolata Schum. 1827: 33 and 1828: 53; Junghans 1962: 97.

Exsicc.: "Guinea", *Thonning* 348 (**C**—1 sheet, type, microf. 1: I. 1, 2); "e Guinea" (**FI**—1 sheet).

"*A. tristachya* Vahl comes very close to our plant, but differs in the stem, which is less tuberous at the base, with stiffer and thicker leaves, ovate more compressed and obtuse spikelets, with pale-white glumes and green keel and awn. Vahl mentions no involucrum universale in *A. tristachya* but it is nevertheless present in the specimen which is in his herbarium."

Fuirena umbellata *Rottb.*—FTA. 8: 467 (1902); FWTA. ed. 2, 3: 325 (1972).

F. pentagona Schumacher 1827: 42 and 1828: 62; Junghans 1962: 99.

Exsicc.: Whydah, 1785, *Isert* s.n. (**C**—1 sheet, microf. 43: II. 1, 2); "Guinea", *Thonning* s.n. (**C**—1 sheet, type, microf. 43: I. 6, 7); "e Guinea" (**P–JU**—1 sheet, No. 2016, microf. 125: III. 1).

"Comes near *F. umbellata* but differs in having thinner culms, narrower leaves, fewer umbels which bear fewer spikes along with more acute spikelets."

Kyllinga erecta *Schumacher* 1827: 42 and 1828: 62; FWTA. ed. 2, 3: 307 (1972); Junghans 1962: 99.

Exsicc.: "Guinea", *Thonning* 379 (**C**—2 sheets, types, microf. 63: I. 4–7); "e Guinea" (**C**—1 sheet, microf. 63: I. 3; **P–JU**—1 sheet, No. 1708, microf. 112. I. 1).

Kyllinga nemoralis *(Forsk.)* *Dandy ex Hutch.*—FWTA. ed. 2, 3: 307 (1972).

Cyperaceae indet. in Junghans 1962: 101.

Exsicc.: "e Guinea" (Hofman Bang) (**C**—1 sheet, microf. 63: II. 1).

Kyllinga squamulata *Vahl* 1805: 381 ("Habitat in Guinea. *Thonning*"); Schum. 1827: 41 and 1828: 61; FWTA. ed. 2, 3: 304 (1972); Junghans 1962: 100.

Exsicc.: "Guinea", *Thonning* 347 (**C**—4 sheets, types, microf. 63: II. 2–7; III. 1–3); "e Guinea" (**P–JU**—1 sheet, No. 1708 bis, microf. 112: I. 1).

"I have seen 12 capitula with three-leaved involucre, and only one with a four-leaved one which Vahl cites in his description."

Lipocarpha sphacelata *(Vahl)* *Nees*—FWTA. ed. 2, 3: 328 (1972). *L. filiformis* (Vahl) Kunth—FTA. 8: 470 (1902).

Hypaelyptum filiforme Vahl 1805: 284 ("Habitat in Guinea. *Thonning*"); Schum. 1827: 31 and 1828: 51; Junghans 1962: 99.

Exsicc.: "Guinea", *Thonning* s.n. (**C**—2 sheets, types, microf. 54: III. 6, 7; 55: I. 1, 2); "e Guinea" (**C**—1 sheet, microf. 54: III. 5; **P–JU**—1 sheet, No. 1781, microf. 116: II. 2).

Mariscus alternifolius *Vahl* 1805: 376 ("Habitat in Guinea. *Thonning*"); Schum. 1827: 40 and 1828: 60; FWTA. ed. 2, 3: 296 (1972); Junghans 1962: 100.

Exsicc.: "Guinea", *Isert* s.n. (**C**—2 sheets, microf. 66: I. 3; II. 1, 2); "Guinea", *Thonning* 384 (**C**—2 sheets, types, microf. 66: I. 4–7); "e Guinea" (Hofman Bang) (**C**—1 sheet, 66: II. 3).

[**Mariscus cyperinus** *(Retz.)* *Vahl* 1805: 377 ("Habitat . . . in Guinea. *Thonning*"); Junghans 1962: 100.

Exsicc.: (no material from Guinea at **C**; see note).

Note: An Indian species: Thonning's specimen has probably been re-determined.—F.N.H.]

Mariscus dubius (Rottb.) C. E. C. Fischer—FWTA. ed. 2, 3: 295 (1972).

Cyperus coloratus Vahl 1805: 312 ("Habitat in Guinea. *Thonning*"); Schum. 1827: 34 and 1828: 54; Junghans 1962: 97. *Mariscus coloratus* (Vahl) Nees—FTA. 8: 381 (1901).

Exsicc.: "Guinea", *Thonning* 396 (**C**—2 sheets, types, microf. 31: II. 4–7).

Mariscus ligularis *(L.) Urb.*—FWTA. ed. 2, 3: 295 (1972). *Mariscus rufus* Kunth—FTA. 8: 396 (1902).

Cyperus ligularis L.—Vahl 1805: 371; Schum. 1827: 40 and 1828: 60; Junghans 1962: 98.

Exsicc.: "Guinea", *Isert* s.n. (**C**—1 sheet, microf. 32: II. 2); "Guinea", *Thonning* 346 (**C**—2 sheets, microf. 32: I. 5–7; II. 1); "e Guinea" (**FI**—1 sheet; **P–JU**—1 sheet, No. 1884 bis, microf. 122. I. 2).

Pycreus polystachyos *(Rottb.) P. Beauv.* FWTA. ed. 2, 3: 301 (1972).

Cyperus angustifolius Schumacher 1827: 38 and 1828: 58; Junghans 1962: 97.

Exsicc.: "Guinea", 1785, *Isert* s.n. (**C**—1 sheet, microf. 31: I. 3, 4); "Guinea", *Thonning* 410 (**C**—1 sheet, type, microf. 31: I. 1, 2).

"At first glance this species much resembles *C. paniculatus*, but it differs in having a more leafy culm, longer leaves, more delicate spikelets and especially in the eight-leaved involucre."

Pycreus pumilus *L.*—FWTA. ed. 2, 3: 302 (1972).

Cyperus patens Vahl 1805: 334 ("Habitat in Guinea. *Thonning*"); Schum. 1827: 36 and 1828: 56; FTA. 8: 295 (1901); Junghans 1962: 98.

Exsicc.: "Guinea", *Thonning* 407 (**C**—2 sheets, types, microf. 33: I. 7; II. 1–3).

Remirea maritima *Aubl.*—Vahl 1805: 391; Schum. 1827: 43 and 1828: 63; FTA. 8: 486 (1902); FWTA. ed. 2, 3: 297 (1972); Junghans 1962: 100.

Exsicc.: "Guinea", *Thonning* 377 (**C**—2 sheets, microf. 91: III. 2–5).

"The Guinea plant differs from that of Cayenne in the larger and more pointed glumes."

Rhynchospora corymbosa *(L.) Britt.*—FWTA. ed. 2, 3: 331 (1972).

Cyperaceae indet. in Junghans 1962: 101.

Exsicc.: Whydah, 1785, *Isert* s.n. (**C**—1 sheet, microf. 94: II. 1, 2).

Rhynchospora holoschoenoides *(L. C. Rich.) Herter*—FWTA. ed. 2, 3: 329 (1972).

Cyperaceae indet. in Junghans 1962: 101.

Exsicc. "Guinea", 1786, *Isert* s.n. (**C**—1 sheet, microf. 94: I. 6, 7).

[**Rhynchospora minutiflora** *(L. C. Rich. ex. Spreng.) Adams* in Phytologia 21, 2: 70 (1971). *R. micrantha Vahl*—FWTA. ed. 2, 3: 333 (1972).

Exsicc.: "Guinea" (see note), 1786, *Isert* s.n. (**C**—1 sheet, microf. 94: II. 5, 6).

Note: Since this American species has not otherwise been recorded from Africa, I suggest that the specimen was really collected in the West Indies where Isert botanised during his homeward journey, thus the American genus *Isertia* Schreb. was named after him. See also note under *Pharus latifolius*, p. 149.—F.N.H.]

Scirpus aureiglumis *Hooper*—FWTA. ed. 2, 3: 310 (1972).

Schoenus junceus Willd. 1794: 2, pl. 1, f. 4. and 1797a: 259 ("Habitat in Guinea."); Junghans 1962: 100.

Exsicc.: "Guinea", *Isert* s.n. (**C**—1 sheet, type, microf. 94: III. 7; 95: I. 1, 2).

Scleria verrucosa *Willd.* 1805: 313 ("Habitat in Guinea."); Schum. 1827: 403 and 1829: 177; FTA. 8: 509 (1902); FWTA. ed. 2, 3: 340 (1972); Junghans 1962: 101.

Exsicc.: Aquapim, 1786, *Isert* s.n. (**B**—1 sheet, No. 17317, microf. 1251: I. 5; **C**—2 sheets, types, microf. 97: II. 6, 7; III. 1, 2).

"In rather damp places at Aquapim."

DIOSCOREACEAE

Dioscorea alata *L.*—Schum. 1827: 447 and 1829: 221; FWTA. ed. 2, 3: 152 (1968); Junghans 1962: 101.

Exsicc.: (no material traced).

"Cultivated".

Dioscorea bulbifera *L.*—FWTA. ed. 2, 3: 152 (1968).

D. sativa sensu Schum. 1827: 447 and 1829: 221, 222; Junghans 1962: 101, non L.

Exsicc.: (no material traced).

"Cultivated".

FLAGELLARIACEAE

Flagellaria guineensis *Schumacher* 1827: 181 and 1828: 201 (no MS); FTA. 8: 90 (1901); FWTA. ed. 2, 3: 51 (1968); Junghans 1962: 101.

Exsicc.: "Guinea", *Isert* s.n. (**C**—1 sheet, microf. 43: I. 1); "Guinea," *Thonning* s.n. (**C**—2 sheets, types, microf. 43: I. 2–5).

"Not unlike *Flagellaria indica* but in this latter, the sheaths are connate, the inflorescences pyramidal-cymose, rachis and flower-stalks branched-paniculate, compressed, the lower two inches in length, rather erect. S."

GRAMINEAE*

Andropogon canaliculatus *Schumacher* 1827: 52 and 1828: 72; FWTA. ed. 2, 3: 486 (1972); Junghans 1962: 102.

* Determinations revised by Dr. C. E. Hubbard and Dr. W. D. Clayton (Kew).

Exsicc.: *Thonning* s.n. (no type material traced).*

"It has so much resemblance to *A. distachyus* that at first glance it is taken for it, but *distachyus* differs from our plant in having larger flowers, downy calyx-segments which are all provided with awns, and with raised streaks and not channelled; the awns on the corolla of the hermaphrodite flower are an inch long and the pedicels of the male flower are narrower, less shiny and not beset with small dots."

Andropogon guineensis *Schumacher* 1827: 51 and 1828: 71; FWTA. ed. 2, 3: 489 (1972); Junghans 1962: 102.

Exsicc.: *Thonning* s.n. (no type material traced).*

"Related to *A. hirtus* but sufficiently differing."

Note: Probably *A. gayanus* Kunth but the description is inconclusive and the name should be rejected—W.D.C.

Andropogon tectorum *Schum. et Thonn.* in Schum. 1827: 49 and 1828: 69; FWTA. ed. 2, 3: 488 (1972); Junghans 1962: 102.

Exsicc.: *Thonning* s.n. (no type material traced).*

"The commonest grass which occupies nearly all fields from the seashore to close up to the mountains."

"The inhabitants make use of it for thatching and in the neighbourhood of Rio Volta they use it for a sort of mat for enclosing their farmsteads."

Anthephora cristata *(Doell)* *Hack. ex De Wild. et Dur.*—FWTA. ed. 2, 3: 457 (1972).

Gramineae indet. in Junghans 1962: 106.

Exsicc.: Whydah, 1785, *Isert* s.n. (**C**—1 sheet, microf. 6: II. 1).

Aristida adscensionis *L.*—FWTA. ed. 2, 3: 379 (1972).

A. submucronata Schumacher 1827: 47 and 1828: 67.

Exsicc.: "Guinea", *Thonning* 356 (**C**—2 sheets, types, microf. 6: II.7; III. 1–3).

A guineensis Trin. et Rupr., Sp. Gram. Stipac. 137 (1842); Junghans 1962: 102.

Exsicc.: *Thonning* (356?) (**LE**—1 sheet, type).

A. thonningii Trin. et Rupr. 1842: 137; Junghans 1962: 103.

Exsicc.: *Thonning* (356?) (**LE**—1 sheet, type).

Aristida caerulescens *Desf.*—Schum. 1827: 47 and 1828: 67; FWTA. ed. 2, 3: 381 (1972); Junghans 1962: 102.

Exsicc.: (no material traced).

* The Andropogoneae have disappeared from Copenhagen and should be looked for in other institutions. There was a possibility that Trinius had them at Leningrad, but a search there has not been successful.

Aristida sieberana *Trin.*—FWTA. ed. 2, 3: 381 (1972).

A. longiflora Schumacher 1827: 48 and 1828: 68; Junghans 1962: 102.

EXSICC.: *Thonning* 357 (**C**—2 sheets, types, microf. 6: III. 4–7).

A. leiocalycina Trin. et Rupr. in Mém. Acad. Pétersb. Sér. 6, 7: 161 (1849).

EXSICC.: "Guinea", *Thonning* (357?) (**LE**—1 sheet, type).

Brachiaria deflexa (*Schumacher*) *C. E. Hubbard*—FWTA. ed. 2, 3: 444 (1972).

Panicum deflexum Schumacher 1827: 63 and 1828: 83; Junghans 1962: 104.

EXSICC.: "Guinea", *Thonning* 390 (**C**—3 sheets, types, microf. 75: II. 3–5).

"Related to P. *glutinosum.*"

P. regulare Nees, Fl. Afr. Austr. 41 (1841).

EXSICC.: "Guinea", *Thonning* s.n. (**S**—1 sheet, type).

Brachiaria distichophylla (*Trin.*) *Stapf*—FWTA. ed. 2, 3: 444 (1972).

Panicum serrulatum Schumacher 1827: 62 and 1828: 82; Junghans 1962: 104.

EXSICC.: "Guinea", *Thonning* 395 (**C**—4 sheets, types, microf. 76: II. 6, 7; III. 1, 2).

P. viviparum Schumacher 1827: 62 and 1828: 82; Junghans 1962: 105.

EXSICC.: "Guinea", *Thonning* 385 (**C**—1 sheet, type, microf. 77: III. 1; **S**—1 sheet, type).

Brachiaria falcifera (*Trin.*) *Stapf*—FWTA. ed. 2, 3: 443 (1972).

Panicum collare Schumacher 1827: 60 and 1828: 80; FTA. 9: 518 (1919); Junghans 1962: 104.

EXSICC.: "Guinea", *Thonning* 373 (**C**—1 sheet, type, microf. 75: II. 2).

"In habit like P. *coloni.* S."

Brachiaria lata (*Schumacher*) *C. E. Hubbard*—FWTA. ed. 2, 3: 444 (1972).

Panicum latum Schumacher 1827: 61 and 1828: 81; Junghans 1963: 104.

EXSICC.: "Guinea", *Thonning* 353 (**C**—2 sheets, types, microf. 75: III. 2, 3; **S**—1 sheet, type).

P. exasperatum Nees ex Steud., Syn. Pl. Glum. 62 (1854).

EXSICC.: "Guinea", *Thonning* 353? (**S**—1 sheet, type).

Brachiaria mutica (*Forsk.*) *Stapf*—FWTA. ed. 2, 3: 443 (1972).

Gramineae indet. in Junghans 1962: 106.

EXSICC.: Ada, 1784, *Isert* s.n. (**C**—1 sheet, microf. 76: I. 1).

Cenchrus biflorus *Roxb.*—FWTA. ed. 2, 3: 464 (1972).

C. barbatus Schumacher 1827: 43 and 1828: 63; FTA. 9: 1079 (1934); Junghans 1962: 103.

Exsicc.: "Guinea", *Thonning* 360 (**C**—1 sheet, type, microf. 18: I. 3, 4).

Chloris pilosa *Schumacher* 1827: 55 and 1828: 75; FWTA. ed. 2, 3: 400 (1972); Junghans 1962: 103.

Exsicc.: "Guinea", *Thonning* 371 (**C**—2 sheets, types, microf. 18: III. 4–7).

Ctenium canescens *Benth.*—FWTA. ed. 2, 3: 399 (1972).

Chloris pulchra Schumacher 1827: 56 and 1828: 76; Junghans 1962: 103.

Exsicc.: "Guinea", *Thonning* 366 (**C**—4 sheets, types, microf. 19: II. 2–7; III. 1).

"The Guinea plant is not at all different from the Indian."

Dactyloctenium aegyptium *(L.) P. Beauv.*—FWTA. ed. 2, 3: 395 (1972).

Chloris guineensis Schumacher 1827: 55 and 1828: 75; Junghans 1962: 103.

Exsicc.: "Guinea", *Thonning* 365 (**C**—4 sheets, types, microf. 18: II. 4–7; III. 1–3).

Digitaria horizontalis *Willd.*—FWTA. ed. 2, 3: 453 (1972).

D. reflexa Schumacher 1827: 44 and 1828: 64; FTA. 9: 479 (1919); Junghans 1962: 103.

Exsicc.: "Guinea", *Thonning* 367 (**C**—1 sheet, type, microf. 36: I. 3).

"Not unlike *D. aegyptiaca* but differs therefrom in the characters cited."

D. nuda Schumacher 1827: 45 and 1828: 65; FTA. 9: 479 (1919); Junghans 1962: 103.

Exsicc.: "Guinea", *Thonning* 367 (**C**—1 sheet, type, microf. 36: I. 1).

"Related to *D. umbrosa*."

NOTE: Both the Thonning sheets bear the same number.—F.N.H.

Digitaria leptorhachis *(Pilger) Stapf*—FWTA. ed. 2, 3: 454 (1972).

Exsicc.: Whydah, 1785, *Isert* s.n. (**C**—1 sheet, microf. 36: I. 2).

Echinochloa pyramidalis *(Lam.) Hitchc. et Chase*—FWTA. ed. 2, 3: 439 (1972).

Gramineae indet. in Junghans 1962: 106.

Exsicc.: Whydah, 1785, *Isert* s.n. (**C**—1 sheet, microf. 37: III. 6).

Eleusine indica *(L.) Gaertn.*—FWTA. ed. 2, 3: 395 (1972).

E. glabra Schumacher 1827: 53 and 1828: 73; Junghans 1962: 103.

Exsicc.: Ada, 1784, *Isert* s.n. (**C**—1 sheet, microf. 47: II. 2); "Guinea", *Thonning* 392 (**C**—1 sheet, type, microf. 47: I. 7; II. 1).

Enteropogon macrostachyus (*Hochst. ex A. Rich.*) *Munro ex Benth.*—FWTA. ed. 2, 3: 402 (1972).

Chloris simplex Schumacher 1827: 54 and 1828: 74; Junghans 1962; 103.

Exsicc.: "Guinea", *Thonning* 338 (**C**—4 sheets, types, microf. 19: I. 1–7; II. 1); "e Guinea" (**P–JU**—1 sheet, No. 2745, microf. 174: III. 6).

"Fairly common in open and dry fields."
"The specimen from India in Vahl's collection differs from our plant only in that the culm is more leafy below, tuberous, the lower sheaths slightly hairy, flowers slightly smaller, the awns shorter."

Eragrostis aspera (*Jacq.*) *Nees*—FWTA. ed. 2, 3: 387 (1972).

Poa hippuris Schumacher 1827: 69 and 1828: 89; Junghans 1962: 105.

Exsicc.: "Guinea", *Thonning* 351 (**C**—2 sheets, types, microf. 86: II. 2–5).

Eragrostis cilianensis (*All.*) *Lut.*—FWTA. ed. 2, 3: 390 (1972).

Poa cachectica Schumacher 1827: 66 and 1828: 86; Junghans 1962: 105.

Exsicc.: "Guinea", *Thonning* s.n. (**C**—2 sheets, types, microf. 86: I. 1–4).

"Very like *Poa unioloides*, but nevertheless different."

Eragrostis ciliaris (*L.*) *R. Br.*—FWTA. ed. 2, 3: 386 (1972).

Poa ciliaris L.—Schum. 1827: 67 and 1828: 87; Junghans 1962: 105.

Exsicc.: "Guinea", *Thonning* 352 (**C**—2 sheets, microf. 86: I. 5–7; II. 1).

Eragrostis linearis (*Schumacher*) *Benth.*—FWTA. ed. 2, 3: 391 (1972).

Poa linearis Schumacher 1827: 67 and 1828: 87; Junghans 1962: 105.

Exsicc.: "Guinea", *Thonning* 382 (**C**—4 sheets, types, microf. 86: II. 6, 7; III. 1–6).

Eragrostis turgida (*Schumacher*) *De Wild.*—FWTA. ed. 2, 3: 387 (1972).

Poa turgida Schumacher 1827: 66 and 1828: 86; Junghans 1962; 106.

Exsicc.: Ada, 1784, *Isert* s.n. (**C**—1 sheet, microf. 87: I. 2); "Guinea", *Thonning* 362 (**C**—6 sheets, types, microf. 86: III. 7; 87: I. 1, 3–7; II. 1–3); "e Guinea" (**P–JU**—1 sheet, No. 2463, microf. 156: I. 1).

Hackelochloa granularis (*L.*) *Kuntze*—FWTA. ed. 2, 3: 505 (1972).

Manisuris granularis (L.) Sw.—Willd. 1806a: 945; Schum. 1827: 65 and 1828: 85; Junghans 1962: 103.

Exsicc.: "Guinea", *Isert* s.n. (**C**—1 sheet, microf. 65: III. 1; "Guinea", *Thonning* 406 (**C**—2 sheets, microf. 65: II. 5, 6; III. 2); "e Guinea" (**C**—1 sheet, microf. 65: II. 7).

"The specimens from Guinea differ from the American ones in that the leaves are longer, the sheaths more hairy, the calyx-segments brown and membranous at the margins and with less hairy margins. Furthermore in the specimens from both places the calyx-segments are lanceolate, especially the outer, and not circular."

Heteropogon contortus *(L.) P. Beauv. ex Roem. et Schult.*—FWTA. ed. 2, 3: 473 (1972).

Andropogon contortum L.—Schum. 1827: 48 and 1828: 68; Junghans 1962: 102.

EXSICC.: *Thonning* 345 (no material traced).

Hyperthelia dissoluta *(Nees ex Steud.) W. D. Clayton*—FWTA. ed. 2, 3: 496 (1972).

Gramineae indet. in Junghans 1962: 106.

EXSICC.: Rio Volta, 1784, *Isert* s.n. (**C**—1 sheet, microf. 55: I. 3).

Imperata cylindrica *(L.) P. Beauv.* var. **africana** *(Anderss.) C. E. Hubbard*— FWTA. ed. 2, 3: 464 (1972).

Gramineae indet. in Junghans 1962: 106.

EXSICC.: Ada, 1784, *Isert* s.n. (**C**—1 sheet, microf. 56: II. 4).

Olyra latifolia *L.*—FWTA. ed. 2, 3: 362 (1972).

O. brevifolia Schumacher 1827: 402 and 1829: 176; Junghans 1962: 104.

EXSICC.: "Guinea", *Thonning* 363 (**C**—2 sheets, types, microf. 72: III. 7; 73: I. 1, 2).

"In shady places."
"*Olyra latifolia* L. differs from our plant in having narrow linear not ovate leaves which are 7 inches long, 1½ inches broad; in that the calyx pods in the female flowers are downy-haired and hairy at margins, lanceolate, with less distinct nerves and shorter point. S."

Oplismenus hirtellus *(L.) P. Beauv.*—FTA. 9: 631 (1920); FWTA. ed. 2, 3: 437 (1972).

Panicum incanum Schumacher 1827: 60 and 1828: 80; Junghans 1962: 104.

EXSICC.: "Guinea", *Thonning* 361 (**C**—2 sheets, types, microf. 75: II. 7; III. 1).

"Related to *P. hirtellum* but differs from it."
"Perhaps not different from *Oplismenus africanus* P. Beauv. Flore d'Oware, tab. 68, fig. 1."

Oryza sativa *L.*—Schum. 1827: 181 and 1828: 201; FWTA. ed. 2, 3: 365 (1972); Junghans 1962: 104.

EXSICC.: (no material traced).

"Cultivated".

Oxytenanthera abyssinica (*A. Rich.*) *Munro*—FWTA. ed. 2, 3: 360 (1972).

Gramineae indet. in Junghans 1962: 106.

Exsicc.: Sawi, 1785, *Isert* s.n. (**C**—1 sheet, microf. 75: I. 1).

Panicum brevifolium *L.*—FWTA. ed. 2, 3: 429 (1972).

P. plantagineum Schumacher 1827: 64 and 1828: 84, non Link; Junghans 1962: 104.

Exsicc.: "Guinea", *Thonning* 341 (**C**—1 sheet, type, microf. 76: II. 5; **S**—1 sheet, type).

"Resembles *P. trichoides* Sw. (*P. brevifolium* Lin.) and *P. latifolium* Lin. but seems to me sufficiently distinct in the characters cited."

"It differs from *P. ovalifolium* Beauv. tab. 110, fig. 1 in the smooth culms and alternating branches of the panicle."

Panicum maximum *Jacq.*—FWTA. ed. 2, 3: 429 (1972).

P. sparsum Schumacher 1827: 64 and 1828: 84; FTA. 9: 738 (1920); Junghans 1962: 105.

Exsicc.: Ada, 1784, *Isert* s.n. (**C**—1 sheet, microf. 76: III. 3); "Guinea", *Thonning* 358 (**C**—3 sheets, types, microf. 76: III. 4–6; **S**—1 sheet, type).

Paspalidium geminatum (*Forsk.*) *Stapf*—FWTA. ed. 2, 3: 440 (1972).

Gramineae indet. in Junghans 1962: 106.

Exsicc.: "Guinea", *Thonning* 354 (**C**—1 sheet, microf. 75: II. 6).

Paspalum orbiculare *Forst.*—FWTA. ed. 2, 3: 446 (1972).

Paspalus (sic) *barbatus* Schumacher 1827: 53 and 1828: 73, non (Trin.) Schult. (1827); Junghans 1962: 105.

Exsicc.: Ada, 1784, *Isert* s.n. (**C**—1 sheet, microf. 77: III. 7); "Guinea", *Thonning* 402 (**C**—1 sheet, type, microf. 77: III. 6).

"Has much resemblance to *P. kora*, but seems to me to differ from it."

Pennisetum americanum (*L.*) *K. Schum*—FWTA. ed. 2, 3: 463 (1972).

Gramineae indet. in Junghans 1962: 106.

Exsicc.: "Guinea" (grown from seed), *Isert* s.n. (**C**—1 sheet, microf. 78: III. 3, 4); "Guinea?", *Ryan* s.n. (**C**—2 sheets, microf. 98: III. 2–5).

Pennisetum polystachion (*L.*) *Schult.*—FTA. 9: 1057 (1934); FWTA. ed. 2, 3: 460 (1972).

Panicum cauda-ratti Schumacher 1827: 59 and 1828: 79; Junghans 1962: 104.

Exsicc.: Whydah, 1785, *Isert* s.n. (**C**—1 sheet, microf. 78: II. 6); "Guinea", *Thonning* 374 (**C**—3 sheets, types, microf. 75: I. 4–7; II. 1).

Pennisetum purpureum *Schumacher* 1827: 44 and 1828: 64; FWTA. ed. 2, 3: 461 (1972); Junghans 1962: 105.

EXSICC.: Whydah, 1785, *Isert* s.n. (**C**—1 sheet, microf. 78: II. 7); "Guinea", *Thonning* 355 (**C**—1 sheet, type, microf. 78: III. 1, 2).

Pennisetum subangustum (*Schumacher*) *Stapf et Hubbard*—FWTA. ed. 2, 3: 461 (1972).

Panicum subangustum Schumacher 1827: 59 and 1828: 79; Junghans 1962: 105.

EXSICC.: "Guinea", *Thonning* 375 (**C**—3 sheets, types, microf. 77: I. 3, 4, 7; II. 1–3); "e Guinea" (**C**—2 sheets, microf. 77: I. 1, 2, 5, 6).

[Pharus latifolius *L.*

Gramineae indet. in Junghans 1962: 106.

EXSICC.: "Guinea", 1786, *Isert* s.n. (**C**—1 sheet, microf. 79: II. 6).

NOTE: An American species not since recorded from Africa. See also note under *Rhynchospora micrantha*, p. 141.—F.N.H.]

Rhynchelytrum repens (*Willd.*) *C. E. Hubbard*—FWTA. ed. 2, 3: 454 (1972).

Saccharum repens Willd. 1797a: 322 ("Habitat in Guinea"); Schum. 1827: 47 and 1828: 67; Junghans 1962: 106.

EXSICC.: Ada, 1784, *Isert* s.n. (**B**—1 sheet, No. 1499, type, microf. 87: III. 7; **C**—1 sheet, type, microf. 94: II. 7; **S**—1 sheet, type); "Guinea", *Thonning* 380 (**C**—3 sheets, microf, 94: III. 1–3); "e Guinea" (**P–JU**—1 sheet, No. 2668, microf. 169: III. 1).

Saccharum officinarum *L.*—Schum. 1827: 47; and 1828: 67; FWTA. ed. 2, 3: 466 (1972); Junghans 1962: 106.

EXSICC.: (no material traced).

"Cultivated as well [as the preceding and following species]."

Saccharum spontaneum *L.* var. **aegyptiacum** (*Willd.*) *Hack.*—FTA. 9: 95 (1917); FWTA. ed. 2, 3: 466 (1972).

S. punctatum Schumacher 1827: 46 and 1828: 66; Junghans 1962: 106.

EXSICC.: *Thonning* s.n. (no type material traced).

"Is cultivated for sake of the sugar."

Schizachyrium sp.

Andropogon simplex Vahl ex Schumacher 1827: 49 and 1828: 69; FWTA. ed. 2, 3: 481 (1972); Junghans 1962: 102.

EXSICC.: *Thonning* s.n. (no type material traced).

NOTE: Probably *Schizachyrium sp.* but the description cannot confidently be equated to any of the West African species of this genus—W. D. Clayton.

Setaria anceps *Stapf ex Massey*—FWTA. ed. 2, 3: 423 (1972).

Gramineae indet. in Junghans 1962: 106.

Exsicc.: Whydah, 1785, *Isert* s.n. (**C**—1 sheet, microf. 98: III. 1).

Setaria barbata (*Lam.*) *Kunth*—FTA. 9: 854 (1930); FWTA. ed. 2, 3: 424 (1972).

Panicum lineatum Schumacher 1827: 61 and 1828: 81; Junghans 1962: 104.

Exsicc.: "Guinea", *Thonning* 386 (**C**—1 sheet, type, microf. 75: III. 4).

Setaria chevalieri *Stapf*—FWTA. ed. 2, 3: 424 (1972).

Exsicc.: "Guinea", *Isert* s.n. (**C**—one leaf on 1 sheet with a flower of *Tacca leontopetaloides* q.v., microf. 64: II. 5).

Setaria longiseta *P. Beauv.*—FTA. 9: 836 (1930); FWTA. ed. 2, 3: 423 (1972).

Panicum longifolium Schumacher 1827: 63 and 1828: 83; Junghans 1962: 104.

Exsicc.: "Guinea", *Thonning* 401 (**C**—2 sheets, types, microf. 75: III. 5–7; **S**—1 sheet, type).

Setaria pallide-fusca (*Schumacher*) *Stapf et C. E. Hubbard*—FWTA. ed. 2, 3: 423 (1972).

Panicum pallide-fuscum Schumacher 1827: 58 and 1828: 78; Junghans 1962: 104.

Exsicc.: "Guinea", *Thonning* 344 (**C**—5 sheets, types, microf. 76: I. 2–7; II. 1–4).

"*Panicum glaucum* comes near this but is different."

Setaria sphacelata (*Schumacher*) *Stapf et C. E. Hubbard ex M. B. Moss*—FTA. 9: 795 (1930); FWTA. ed. 2, 3: 423 (1972).

Panicum sphacelatum Schumacher 1827: 58 and 1828: 78; Junghans 1962: 105.

Exsicc.: "Guinea", *Thonning* s.n. (**C**—1 sheet, type, microf. 77: II. 4, 5); "e Guinea" (**C**—1 sheet, microf. 77: II. 6, 7).

Sorghum arundinaceum (*Desv.*) *Stapf*—FWTA. ed. 2, 3: 467 (1972).

Andropogon arundinaceus Willd. 1806a: 906 ("Habitat in Guinea. *Isert*"), non Berg.; Junghans 1962: 102.

Exsicc.: *Isert* s.n. (**B**—1 sheet, No. 18639, type, microf. 1354: III. 6).

Sorghum bicolor (*L.*) *Moench*—FWTA. ed. 2, 3: 467 (1972).

Sorghum vulgare Pers.—Schum. 1827: 57 and 1828: 77; Junghans 1962: 106.

Exsicc.: "Hort. Hafn." (**C**—3 sheets, microf. 103: III. 6, 7; 104: I. 1–4).

"Is cultivated."

Sporobolus pyramidalis *P. Beauv.*—FWTA. ed. 2, 3: 408 (1972).

Agrostis extensa Schumacher 1827: 45 and 1828: 65; Junghans 1962: 102.

EXSICC.: "Guinea", *Thonning* 381 (C—5 sheets, types, microf. 4: I. 3–7; II. 1–5).

"*A. tenacissima* differs from our plant especially in that the opening of the sheath is naked, the panicle filiform and narrow and the leaves of the calyx of same length as those of the corolla".

Sporobolus virginicus *(L.) Kunth*—FWTA. ed. 2, 3: 408 (1972).

Agrostis congener Schumacher 1827: 46 and 1828: 66; Junghans 1962: 101.

EXSICC.: "Guinea", *Thonning* 343 (C—5 sheets, types, microf. 3: III. 1–7; 4: I. 1, 2).

"Intermediate between *A. pungens* and *A. virginica*."

Tricholaena monachne *(Trin.) Stapf et C. E. Hubbard*—FWTA. ed. 2, 3: 455 (1972).

Aira bicolor Schumacher 1827: 65 and 1828: 85; FTA. 10: 88 (1937); Junghans 1962: 102.

EXSICC.: "Guinea", *Thonning* 364 (C—1 sheet, type, microf. 4: II. 6).

Vetiveria fulvibarbis *(Trin.) Stapf*—FTA. 9: 158 (1971); FWTA. ed. 2, 3: 470 (1972).

Andropogon verticillatus Schumacher 1827: 50 and 1828: 70 (no MS or No.), non Roxb.; Junghans 1962: 102.

EXSICC.: *Thonning*? (no type material traced).

Vetiveria nigritana *(Benth.) Stapf*—FWTA. ed. 2, 3: 470 (1972).

Gramineae indet. in Junghans 1962: 106.

EXSICC.: Ada, 1784, *Isert* s.n. (C—1 sheet, microf. 109: II. 5).

Zea mays *L.*—Schum. 1827: 402 and 1829: 176; FWTA. ed. 2, 3: 511 (1972); Junghans 1962: 106.

EXSICC.: No material traced.

"Cultivated".

HYPOXIDACEAE

Curculigo pilosa *(Schum. et Thonn.) Engl.*—FTA. 7: 383 (1898); FWTA. ed. 2, 3: 174 (1968).

Gethyllis pilosa Schum. et Thonn. in Schum. 1827: 172 and 1828: 192; Junghans 1962: 106.

Exsicc.: Accra, 1786, *Isert* s.n. (**C**—1 sheet, microf. 44: III. 3, 4); "Guinea", *Thonning* 9 (**C**—6 sheets, types, microf. 44: II. 2–7; III. 1, 2, one of the sheets not photographed; **S**—1 sheet, type).

"Common in coastal areas, flowers in May and June."

"In Vahl's herbarium lies a plant under the name of *Curculigo orchioides* sent from East Indies by Röttler, which, as regards the leaves, much resembles our plant; but they are somewhat shorter, the scapes are similarly shorter (about 1½ inches) and the corollas devoid of tubes. A second plant in the same herbarium under name of *Hypoxis scorzonerifolia*, sent by v. Rohr from the island of Trinidad, similarly differs in its shorter slightly broader leaves (breadth is nearly an inch) and a scape of 1½ inches in length, very stiff haired, one flowered and a corolla without a tube."

"Owing to the corolla's very long tube and the fleshy-leathery fruit, this plant is best placed in the genus *Gethyllis*. S."

LILIACEAE

Allium guineense *Thonning* in Schum. 1827: 171 and 1828: 191; Junghans 1962: 106.

Exsicc.: *Thonning* 312 (no type material traced).

Vernacular name: 'Sabullá'.

"It is cultivated mostly by the Gah natives."
"The bulb is just as good as a shallot-bulb, but it is somewhat smaller and therefore more delicate. Th."

Note: No doubt this *is* the shallot, *A. ascalonicum* L., which is a well known crop in Ghana (see J. M. Grove in Bull. IFAN. sér. B, 28: 390–404 (1966))—F.N.H.

Aloe picta *Thunb.*—Schum. 1827: 176 and 1828: 196; Junghans 1962: 107.

Exsicc.: Blegusso, *Thonning* 349 (no material traced).

Vernacular name: 'Ahablobae' or 'Asablobae'.

"Rare near the coast, more frequent at Blegusso and other places with a similar situation."
"The leaves contain a clear yellowish mucilaginous bitter juice, which is soluble in water without making it turbid and on being exposed to the air gives the water gradually a stronger and stronger dark red colour. The natives use the crushed leaves for old leg bone injuries, which they likewise wash with a decoction. In dropsy the same concoction is used as a cathartic purge."

Note: It is difficult to know which species is referred to in the absence of specimens. Two varieties, α *major* and β *minor* are given by Schumacher. The following species are recorded from Ghana in FWTA. ed. 2, 3: 92 (1968): *A. schweinfurthii* Bak. (1882), *A. buettneri* A. Berger (1905), *A. keayi* Reynolds (1963). The first and third occur at the present time on the Accra Plains.—F.N.H.

Asparagus warneckei (*Engl.*) *Hutch.*—FWTA. ed. 2, 3: 93 (1968).

Liliaceae indet. in Junghans 1962: 107.

Exsicc.: "Guinea", Hb. Horneman prob. *Thonning* (**C**—1 sheet, microf. 8: II. 1, 2).

Chlorophytum inornatum *Ker-Gawl.*—FWTA. ed. 2, 3: 100 (1968).

Exsicc.: Aquapim, 1786, *Isert* s.n. (**C**—1 sheet, microf. 64: II. 6, 7).

Eucomis undulata *Ait.*, Hortus Kewensis 1: 433 (1789).

Exsicc.: "Guinea?", 1783, *Isert* s.n. (**C**—1 sheet, not photographed).

NOTE: It is surprising to find this South African plant in Isert's collection. In his handwriting there is on the reverse of the sheet "Fritillaria regia L. l. h. h. 1783", with the later addition in another hand of "Eucomis undulata Ait. leg. Isert". Since Isert was in West Africa in 1783 one must conclude that l. h. h. refers to its occurrence in gardens.—F.N.H.

Gloriosa superba *L.*—FTA. 7: 563 (1898); FWTA. ed. 2, 3: 106 (1968).

G. angulata Schumacher 1827: 171 and 1828: 191; Junghans 1962: 107.

Exsicc.: "Guinea", *Thonning* s.n. (**C**—1 sheet, microf. 45: I. 3).

"Resembles *G. superba* but differs in its two-edged, above sinuous stem; leaves which are broader at the base and less veined."

Iphigenia ledermannii *Engl. et K. Krause*—Hepper in Kew Bull. 21: 497 (1968); FWTA. ed. 2, 3: 106 (1968).

Helonias guineensis Thonning in Schum. 1827: 182 and 1828: 202; Junghans 1962: 107, non *Iphigenia guineensis* Baker.

Exsicc.: "Guinea", *Thonning* 106 (**C**—1 sheet, type, microf. 52: III. 1; 4 sheets not photographed; **P–JU**—1 sheet, No. 3244, microf. 213: III. 1).

"Rare, flowers in June."

Urginea ensifolia (*Thonning*) *Hepper* in Kew Bull. 21: 497 (1968); FWTA. ed. 2, 3: 103 (1968).

Ornithogalum ensifolium Thonning in Schum. 1827: 173 and 1828: 193; Junghans 1962: 107.

Exsicc.: "Guinea", *Thonning* 279 (**C**—2 sheets, types, microf. 73: II. 5–7; III. 1).

"In coastal areas, not common; flowers in April."

MARANTACEAE

Marantochloa leucantha (*K. Schum.*) *Milne-Redh.*—FWTA. ed. 2, 3: 83 (1968).

Marantaceae indet. in Junghans 1962: 107.

Exsicc.: Aquapim 1786, *Isert* s.n. (**C**—1 sheet, microf. 65: III. 3).

Marantochloa purpurea (*Ridl.*) *Milne-Redh.*—FWTA. ed. 2, 3: 83 (1968).

Marantaceae indet. in Junghans 1962: 107.

Exsicc.: "Guinea", *Thonning* s.n. (**C**—3 sheets, microf. 65: III. 4–7; 66: I. 1–2).

ORCHIDACEAE

Eulophia cristata (*Sw.*) *Steud.*—FWTA. ed. 2, 3: 249 (1968). *Lissochilus purpuratus* Lindl.—FTA. 7: 79 (1897).

Limodorum articulatum Schum. et Thonn. in Schum. 1827: 399 and 1829: 173; Junghans 1962: 107.

Exsicc.: *Thonning* 256 (no type material traced).

Vernacular name: 'Jangkosno'.

  "Here and there."

Eulophia gracilis *Lindl.*—FTA. 7: 51 (1897); FWTA. ed. 2, 3: 247 (1968).

Limodorum ciliatum Schum. et Thonn. in Schum. 1827: 400 and 1829: 174; Junghans 1962: 107.

Exsicc.: "Guinea", *Thonning* 261 (**C**—2 sheets, type, microf. 64: III. 1–4).

  "Here and there."

Habenaria filicornis (*Thonning*) *Lindl.*—FTA. 7: 216 (1898); FWTA. ed. 2, 3: 193 (1968).

Orchis filicornis Thonning in Schum. 1827: 397 and 1829: 171, non L.f.; Junghans 1962: 107.

Exsicc.: "Guinea", *Thonning* 58 (**C**—4 sheets, types, microf. 73: I. 4–7; II. 1–4).

  "In damp places. Flowers in July."
  "I have some doubt whether the plant should be referred to the genus *Orchis*. S."

PALMAE

Ancistrophyllum secundiflorum (*P. Beauv.*) *Wendl.*—FWTA. ed. 2, 3: 167 (1968).

Palmae indet. in Junghans 1962: 108.

Exsicc.: Fida, 1785, *Isert* s.n. (**C**—3 sheets, microf. 6: I. 5–7).

Borassus aethiopum *Mart.*—FWTA. ed. 2, 3: 168 (1968).

B. flabelliformis Murr.—Schum. 1827: 443 and 1829: 217; Junghans 1962: 107.

Exsicc.: *Thonning* 258 (no material traced).

Vernacular name: 'Vjye-tjo'.

"Here and there."

"The trunk yields the best and most durable beams, but as these are rather short they are not usable for every purpose, moreover they do not attain a symmetrical shape. The trunk is cleft in four parts and the inner fibrous substance removed, it is a pity that convenient tools for felling, cleaving and transporting them are not at hand which make them as expensive as beams from Europe. The young shoot, just sprouting from the soil, is boiled and eaten. The fruit is eaten raw as well as boiled and the natives in particular soak the pulpy substance which is boiled together with roasted and grated maize."

"The gelatinous pips of the unripe fruit are greedily eaten and regarded as very nutritious and as an aphrodisiac."

"From the leaves are made fly-fans, mats, mat-bags etc."

Elaeis guineensis *Jacq.*—Schum. 1827: 439 and 1829: 213; Junghans 1962: 107.

Exsicc.: *Thonning* 333 (no material traced).

Vernacular name: 'Taehn-tio'.

"Grows wild and is cultivated."

"The oil palm is cultivated for its important uses. Hardly any part of the whole palm is without some utility. The fruit is collected and beaten with mallets until the pulpy substance is completely separated from the stones, thereafter the entire dough is soaked in warm water and worked well through with the hands so that some of the oil becomes separated out and still more after being boiled. At Poppo a lamp oil is prepared from the kernels; in Aquapim the stones are burnt along with the kernels to ashes from which is prepared a lye which is boiled with palm oil to make soap. Mats are plaited with the foliage for thatching; this thatch is stronger than common straw-thatch but not so durable. Also mats of it are plaited for fences around farms and houses. The wool which occurs on the stipe is scraped off, mixed with powder and used as tinder under the name of Asoso. The stipes are cut off the 6–10 yr. old palm and the palm is dug up and thrown over; it lies like that for four weeks; after this a four-sided cavity is cut in the middle of the trunk on the upper side, which goes somewhat deeper down into the middle. This opening must be protected with palm leaves from the sun and dust; and in order that the wine which exudes from both ends may have a free course, a hole is bored through the trunk in which a tube is placed and below this a pot. The hole from which the wine flows out must be burnt out every day and freshly cut out. The first wine is the sweetest. Vinegar is often prepared from it. It may cause diarrhoea; that which follows is more fermented; the final lot is acid and bad. A palm may continue to yield wine for 6 weeks. After some time some large thick white worms collect in the hole. These are greedily eaten by the natives (Akon Krong). The palm wine begins to ferment immediately on being tapped and is very soon alcoholic and then acid. It effervesces like seltzer water or champagne."

"The commonest method of preparing the oil is by collecting the ripe fruits and laying them by till they begin to decay. They are then beaten in a little groove which is lined with flat stones; during beating warm water is poured on them and when this has continued long enough, the dough is collected to the sides of the groove in which the oil flows down towards the middle of the same where it is deepest. Then the dough is squeezed with the hands and at last the final oil is boiled out.

"In the tapped off palm trees there frequently occurs a weevil (*Curculio* sp.) which is roasted and eaten by the natives. Th."

Hyphaene guineensis *Schum. et Thonn.* in Schum. 1827: 445 and 1829: 219; FTA. 8: 120 (1901); Junghans 1962: 107; Furtado in Gard. Bull. Singapore 25: 311–334 (1970). *H. thebaica* sensu FWTA. ed. 2, 3: 169 (1968), non (L.) Mart.

EXSICC.: *Thonning* 332 (no type material traced).

VERNACULAR NAME: 'Songu-tjo'.

"On comparing Thonning's descriptions of the palms here given which are made at the place of growth from fresh specimens, with those of Jacquin, Wildenow, Gaertner and Persoon, important divergences will be found. S."

NOTE: Furtado (*op. cit.* 331) makes a neotype of a specimen collected in Ghana now at Kew labelled "Colonial & Indian Exhibition of 1886 No. 2". —F.N.H.

Phoenix reclinata *Jacq.*—FTA. 8: 103 (1901); FWTA. ed. 2, 3: 169 (1968).

P. spinosa Schum. et Thonn. in Schum. 1827: 437 and 1829: 211; Junghans 1962: 108.

EXSICC.: "Guinea", 1785, *Isert* s.n. (**C**—1 sheet, microf. 80: II. 6, 7); R. Volta, Aquapim etc., *Thonning* 101 (**C**—5 sheets, types, microf. 80: I. 5–7; II. 1–5; III. 1).

VERNACULAR NAME: 'Akoteno'; fruit 'Amitjolobi'; 'Uaegte vin palme', 'Söd vin palme' (Danish).

"At Rio Volta, on the Aquapim mountains and some other places."

"The juice of this palm is far sweeter, but less fermented than that of the true wine-palm (*Elais guineensis*) but where this latter cannot be had one has to be satisfied with the former. They tap the sap in the following way: when the tree has reached a man's height or somewhat over, the leaves are cut off close to the trunk; about the 8th day after, the top is cut off and the sap which therefore wells out is conducted in a curved tube to a bottle or calabash which is tightly fixed to the trunk; a new cut must be made every day as the old orifices get stopped up; when the tree will yield no more sap of itself, a fire is lighted at the lowest part of the trunk and this drives the remaining sap upwards and completely exhausts the tree. Most natives on the Rio Volta suffer much from tumours (hydrocele) which are often extremely large; it is attributed to this drink but might as well be caused by their low-lying and damp country which at one time of year is almost entirely flooded, or to their immoderate pytto drinking (many natives may drink 16–24 pots of pytto a day). Cords are plaited with the young leaves and they prove fairly strong. The fruit has a very light and sweetish pulp; it is called Amitjolobi."

TACCACEAE

Tacca leontopetaloides (*L.*) *Kuntze*—FWTA. ed. 2, 3: 176 (1968).

T. involucrata Schum. et Thonn. in Schum. 1827: 177 and 1828: 197; Junghans 1962: 108.

EXSICC.: "Guinea", 1786, *Isert* s.n. (**C**—1 sheet, microf. 106: I. 5); "Guinea", *Thonning* 293 (**C**—1 sheet, type, microf. 106: I. 3, 4).

"In the less wooded field areas, not very common."
"Although Thonning's description especially as regards absence of a calyx differs from that of others, I have nevertheless no doubt that the plant belongs to this genus; but on account of the involucre and the single scape I regard it as a separate species, for the authors make no mention of any involucre in *T. pinnatifida*; and Ammann in Comment. Petropolit. VIII p. 211 (*Leontopetaloides*) ascribes 2 scapes to it, similarly also his fig. 13 shows that the leaves are not tripartite but bipartite. S."
"The root is not eaten by the natives."

Liliaceae indet. in Junghans 1962: 106.

EXSICC.: "Planta mihi adhoc omnino ignota. hab. in umbros. sylvaticis montosis." "Guineae", *Isert* s.n. (**C**—one flower on one sheet with leaf of *Setaria chevalieri* q.v., microf. 64: II. 5).

TYPHACEAE

Typha domingensis *Pers.*—FTEA. Typhac. 2 (1971).

T. australis Schum. et Thonn. in Schum. 1827: 401 and 1829: 175; FTA. 8: 135 (1901); FWTA. ed. 2, 3: 131 (1968); Junghans 1962: 108.

EXSICC.: Quitta, *Thonning* 341 (**C**—1 sheet, type, microf. 107: III. 4, 5).

VERNACULAR NAME: 'Kaasaamae'.

"At Quitta."
"Much resembles the other species, but it differs from *T. latifolia* and *angustifolia* in its interrupted spike. Wonder if different from *T. domingensis* Pers. Syn. 2. p. 5?"

ZINGIBERACEAE

Aframomum melegueta *K. Schum.*—FWTA. ed. 2, 3: 76 (1968).

Amomum granum-paradisi sensu Schum. 1827: 396 and 1829: 170; Junghans 1962: 108, non L.

EXSICC.: Aquapim, 1786, *Isert* s.n. (**C**—1 sheet, microf. 5: II. 6).

"In shady places in Aquapim."

"The seeds are used in different ways by the inhabitants, partly in economic, partly in medicinal respect."

Aframomum sceptrum *(Oliv. et Hanb.) K. Schum.*—FWTA. ed. 2, 3: 76 (1968).

Exsicc.: "Guinea", *Thonning* s.n. (**C**—3 sheets, microf. 5: I. 7; II. 1–5).

Zingiberaceae indet. in Junghans 1962: 108.

Exsicc.: Whydah, 1785, *Isert* s.n. (**C**—1 sheet, microf. 5: I. 6).

Costus afer *Ker-Gawl.*—FWTA. ed. 2, 3: 78 (1968).

C. arabicus sensu Schum. 1827: 396 and 1829: 170; Junghans 1962: 108, non L.

Exsicc.: Aquapim, *Thonning* s.n. (**C**—1 sheet, microf. 27: I. 6, 7).

"Here and there in Aquapim."

Zingiberaceae indet. in Junghans 1962: 108.

Exsicc.: Aquapim, 1786, *Isert* s.n. (**C**—1 sheet, microf. 5: I. 5).

Curcuma domestica *Valeton*—FWTA. ed. 2, 3: 70 (1968).

C. longa sensu Schum. 1827: 397 and 1829: 171; FWTA. ed. 2, 3: 70 (1968); Junghans 1962: 108, non L.

Exsicc.: *Thonning*? (no material traced).

"Cultivated".

Zingiber officinale *Rosc.*—FWTA. ed. 2, 3: 70 (1968).

Amomum zingiber L.—Schum. 1827: 395 and 1829: 169; Junghans 1962: 108.

Exsicc.: *Thonning*? (no material traced).

"Cultivated".

PTERIDOPHYTES*

Acrostichum aureum *L.*—Alston 1959: 36.

Pteridophyta indet. in Junghans 1962: 110.

Exsicc.: "Guinea", 1784, *Isert* s.n. (**C**—1 sheet, microf. 2: II. 6).

Arthropteris orientalis *(Gmel.) Posth.*—Alston 1959: 52.

Aspidium thonningii Schumacher 1827: 455 and 1829: 299 (no MS).

* Determinations according to Alston (1959) and others revised by Dr. F. M. Jarrett (Kew).

Exsicc.: Aquapim, 1786, *Isert* s.n. (**C**—1 sheet, type, microf. 8: III. 4).

"Grows at Aquapim."

Arthropteris palisotii *Desv.*—Alston 1959: 52.

Adiantum sublobatum Schumacher 1827: 461 and 1829: 235 (no MS); Junghans 1962: 109.

Exsicc.: Aquapim, *Thonning* 308 (**C**—1 sheet, type, microf. 3: I. 7; II. 1); "e Guinea" (**C**—1 sheet, microf. 3: I. 6).

"With the foregoing species" (i.e. *Adiantum palmatum*).
"In the absence of sori it is uncertain whether it belongs here or to *Trichomanes* or *Hymenophyllum*. S."

Pteridophyta indet. in Junghans 1962: 110.

Exsicc.: "e Guinea", *Mortensen* s.n. (**C**—2 sheets, microf. 7: I. 1; 89: I. 3, 4).

Asplenium africanum *Desv.*—Alston 1959: 55.

A. guineense Schumacher 1827: 458 and 1829: 232 (no MS).

Exsicc.: Aquapim, *Thonning* s.n. (**C**—1 sheet, type, microf. 10: I. 4).

"In Aquapim."
"Seems to me to be different from *Asplenium serratum* which it much resembles. In the latter the stipe is shorter, the leaf longer and broader, the tip more obtuse but with a drawn-out point; sori are more frequent and closer together. S."

Azolla africana *Desv.*-Alston 1959: 26 (fig. 7), 27.

A. guineensis Schumacher 1827: 462 and 1829: 236 (no MS).

Exsicc.: "Guinea", *Thonning* s.n. (**C**—1 sheet, microf. 10: II. 1, 2).

"In fresh water."

Blechnum seminudum *Willd.* 1794: 13, pl. 8 f. 2 ("Habitat in Guinea. *Isert*").

Grammitis seminuda (Willd.) Willd. 1810: 140.

Exsicc.: *Isert* (**B**—1 sheet, No. 19587, type, microf. 1416: II. 1).

Ctenitis cirrhosa (*Schumacher*) *Ching*—Alston 1959: 71.

Aspidium cirrhosum Schumacher 1827: 457 and 1829: 231 (no MS).

Exsicc.: Aquapim, *Thonning* 302 (**C**—3 sheets, types, microf. 8: II. 3–5).

"In Aquapim."

Ctenitis protensa (*Afzel. ex Sw.*) *Ching*—Alston 1959: 71, 72 (fig. 14, C, D).

Polypodium pubescens Schumacher 1827: 453 and 1829: 227 (no MS).

Exsicc.: Aquapim, *Thonning* 301 (**C**—1 sheet, type, microf. 89: I. 2).

"Grows at Aquapim."

Cyclosorus dentatus *(Forsk.) Ching*—Alston 1959: 62.

Aspidium aquapimense Schumacher 1827: 456 and 1829: 230 (no MS).

Exsicc.: Aquapim, 1786, *Isert* s.n. (**C**—1 sheet, type, microf. 9: II. 1); "Guinea", *Mortensen* s.n. (**C**—4 sheets, microf. 9: I. 4–7); "e Guinea" (**C**—7 sheets, microf. 8: III. 5–7; 9: I. 1–3; II. 2).

"Grows at Aquapim."
"Has much resemblance to *Aspidium rotundatum* but appears to be different from it. S."

NOTE: Alston (1959: 63) included *Aspidium aquapimense* with a ? under *Cyclosorus afer*. Now R. E. Holttum has kindly confirmed that it is a synonym of *C. dentatus*.—F.N.H.

Cyclosorus striatus *(Schumacher) Ching*—Alston 1959: 62.

Aspidium striatum Schumacher 1827: 456 and 1829: 230 (no MS).

Exsicc.: Whydah, 1784, *Isert* s.n. (**C**—1 sheet, type, microf. 8: III. 3).

"In damp places at Whydah and Aquapim."

Diplazium proliferum *(Lam.) Kaulf.*—Alston 1959: 65; Junghans 1962: 110.

D. incisum Schumacher 1827: 458 and 1829: 232.

Exsicc.: see below.

"In shady places in Aquapim."
"Much like *Diplazium grandifolium* but in the latter the pinnae are the same length from base to tip, serrately-toothed not incised, almost two-lobed at the base, the lowest lobe longer, rounded, the upper-most one smaller. S."

D. serratum Schumacher 1827: 459 and 1829: 233.

"Grows with the foregoing species" [*D. incisum*].
"I dare not decide whether it is a separate species or a subspecies of the foregoing, but it is sufficiently distinct from all the other species. S."

Exsicc.: Aquapim, 1786, *Isert* s.n. (**C**—1 sheet, microf. 36: III. 7); "Guinea", *Thonning* s.n. (**C**—3 sheets, types, microf. 36: II. 5–7; III. 1, 2); "e Guinea" (**C**—3 sheets, microf. 36: II. 2, 3; III. 3, 4, 6); "Guinea", *Mortensen* s.n. (**C**—1 sheet, microf. 36: III. 5).

NOTE: All the specimens of Schumacher's two species have been cited together.—F.N.H.

Doryopteris kirkii *(Hook.) Alston*—Alston 1959: 43.

Adiantum palmatum Schumacher 1827: 460 and 1829: 234 (no MS).

Exsicc.: Aquapim, *Thonning* 307 (**C**—2 sheets, types, microf. 3: I. 1–4);
"e Guinea" (**C**—3 sheets, microf. 2: III. 4–7; 3: I. 5).

"In Aquapim."
"The leaf is the same shape as *Pteris pedata*, but the pinnae are deeper and
more regularly incised."

Lygodium microphyllum *(Cav.) R. Br.*—Alston 1959: 22.

Pteridophyta indet. in Junghans 1962: 110.

Exsicc.: Whydah, 1785, *Isert* s.n. (**C**—1 sheet, microf. nil).

Marsilea minuta *L.*—Launert in Kew Bull. 28: 319 (1973).

M. fimbriata Schumacher 1827: 461, and 1829: 235 (no MS.); Alston 1959:
25.

Exsicc.: "Guinea", 1787, *Isert* s.n. (**C**—1 sheet, type, microf. 66: II. 4).

"In fresh-water swamps."
"Comes near *Marsilea quadrifolia* but in the latter the pinnae are smooth,
and the stalks shorter."

Microlepia speluncae *(L.) Moore*—Alston 1959: 33.

Pteridophyta indet. in Junghans 1962: 110.

Exsicc.: "e Guinea", *Mortensen* s.n. (**C**—2 sheets, microf. 67: I. 7; II. 1–3).

Microsorium punctatum *(L.) Copel.*—Alston 1959: 49.

Polypodium crassinerve Schumacher 1827: 453 and 1829: 227 (no MS).

Exsicc.: Aquapim, 1786, *Isert* s.n. (**C**—1 sheet, type, microf. 89: I. 1).

"Grows at Aquapim."
"The leaf-shape resembles *Polypodium crassifolium*, but in the latter the leaf
is somewhat undulate, less narrow at the base, and the sori much larger and
arranged in rows."

Nephrolepis biserrata *(Sw.) Schott*—Alston 1959: 50.

Aspidium guineense Schumacher 1827: 455 and 1829: 229 (no MS); Junghans
1962: 109.

Exsicc.: Aquapim, *Thonning* s.n. (**C**—2 sheets, types, microf. 8: II. 6, 7).

"Grows at Aquapim."
"Closely approaches *Aspidium punctulatum*, but in the latter the midmost
pinnae are longest, the upper and lower shorter, the serrations are simple,
not split; sori smaller and further from the margin."

A. punctulatum sensu Schum. 1827: 454 and 1829: 228 (no MS), non Sw.;
Junghans 1962: 109.

Exsicc.: Whyda, 1785, *Isert* s.n. (**C**—1 sheet, microf. 10: I. 3); "Guinea",
Mortensen s.n. (**C**—2 sheets, microf. 9: III. 4–7); "e Guinea" (**C**—4 sheets,
microf. 9: II. 5–7; III. 1–3; 10: I. 1, 2).

"Grows at Whyda."

"Related to *Aspidium splendens* & *A. acuminatum* but sufficiently different from them. S."

Pteridophyta indet. in Junghans 1962: 110.

Exsicc.: "e Guinea" (**C**—1 sheet, microf. 9: II. 3, 4).

Ophioglossum costatum *R. Br.*—Alston 1959: 18.

O. fibrosum Schumacher 1827: 452 and 1829: 226; Junghans 1962: 110.

Exsicc.: Frederiksberg, *Thonning* 50 (**C**—1 sheet, type, microf. 73: I. 3).

"Grows in the rainy season in the valley behind the plantation of Frederiksberg."

"Related species are *Ophioglossum vulgare, ovatum* and *bulbosum,* but all differ from our plant. In *O. vulgatum* the leaf is more ovate, obtuse, the root consists of a few thick brown fibres; in *O. ovatum* the root has single fibre; in *O. bulbosum* the whole plant is much smaller, the leaf ovate-cordate, the root tuberous, not fibrous. S."

Pteris atrovirens *Willd.*—Alston 1959: 42.

P. spinulifera Schumacher 1827: 459 and 1829: 233 (no MS); Junghans 1962: 110, partly.

Exsicc.: "Guinea", *Mortensen* s.n. (**C**—2 sheets, microf. 90: II. 3, 4); "e Guinea" (**C**—5 sheets, types, microf. 90: II. 2, 6, 7; III. 1, 2).

"In Aquapim."

"Related species are *Pteris atrovirens* and *P. nemorosa.* S."

Note: See also *P. acanthoneura*—F.N.H.

Pteris acanthoneura *Alston*—Alston 1959: 42.

P. spinulifera Schumacher—Junghans 1962: 110, partly.

Exsicc.: Aquapim, *Thonning* 305 (**C**—1 sheet, type, microf. 90: II. 5); "e Guinea" (**C**—1 sheet, microf. 90: II. 1).

Note: See notes under *P. atrovirens*—F.N.H.

Pteris togoensis *Heiron.*—Alston 1959: 40.

Pteridophyta indet. in Junghans 1962: 110.

Exsicc.: "e Guinea", *Mortensen* s.n. (**C**—1 sheet, microf. 90: I. 7).

Selaginella myosurus *(Sw.) Alston*—Alston 1959: 15.

Pteridophyta indet. in Junghans 1962: 110.

Exsicc.: "Guinea", *Mortensen* s.n. (**C**—1 sheet, microf. 98: III. 6; 99: I. 3–4); "e Guinea" (**C**—2 sheets, microf. 98: III. 7; 99: I. 1–2).

Tectaria angelicifolia (*Schumacher*) *Copel.*—Alston 1959: 74.

Polypodium angelicaefolium Schumacher 1827: 454 and 1829: 228 (no MS).

Exsicc.: Aquapim, *Isert* s.n. (**C**—1 sheet, microf. 88: III. 6); "Guinea", *Thonning* 304 (**C**—1 sheet, type, microf. 88: III. 7).

"Grows near Aquapim."

Thelypteris elata (*Mett. ex Kuhn*) *Schelpe* in Journ. S. Afr. Bot. 31: 265 (1965).

Cyclosorus elatus (Mett. ex Kuhn) Alston—Alston 1959: 25.

Exsicc.: "Guinea", *Thonning* s.n. (**C**—1 sheet, microf. nil).

FUNGI

Laschia delicata *Fr.* in Acta R. Soc. Sci. Upsala ser. 3, 1: 105 (1851).

Excicc.: "ad truncos in Guinea, *Isert*" (no material traced).

REFERENCES

This list supplements those given in the text. For the most part the older references or those frequently cited in the text have been abbreviated and are listed here. Additional references may be traced through the appropriate volume of the Flora of West Tropical Africa, Revised Edition, and in Junghans (1961 & 1962).

Alston, A., 1959: Ferns and fern-allies of West Tropical Africa.

Baillon, H. E., 1860: Species *Euphorbiacearum*, *Euphorbiacées* africaines I in Adansonia 1: 58–87.

Beauvois, Palisot de, 1804: Flore d'Oware et de Benin, en Afrique 1, 1.

Cambessedes, J., 1829: Mémoire sur la famille des *Sapindacées*.—Mém. Mus. d'Hist. Nat. Paris 18: 1–50, Pl. 1–3.

De Candolle, A., 1844: Prodromus 8.
  —1845: Prodromus 9.
  —1847: Prodromus 11.

De Candolle, A. P., 1818: Regni vegetabilis systema naturale 1.
  —1824: Prodromus 1.
  —1825: Prodromus 2.
  —1825–26: Plantes rares du Jardin de Genève.
  —1828: Prodromus 3.
  —1830: Prodromus 4.

Didrichsen, F., 1855: Plantes nonnullas musei Universitatis Hauniensis descripsit.— Vidensk. Medd. naturhist. Forening i Kjöbenhavn, for Aaret 1854: 182–200.

Don, G., 1832: A general system of gardening and botany 2.

Dunal, M.-F., 1817: Monographie de la famille des *Anonacées*.

Fox Maule, A. 1974: Danish botanical expeditions and collections in foreign continents. Botanisk Tidsskrift 69: 167–205.

FTA. vol. and date as indicated: Flora of Tropical Africa 1868–1937.

FTEA. family and date as indicated: Flora of Tropical East Africa 1952 onwards.

FWTA. vol. and date as indicated: J. Hutchinson & J. M. Dalziel, Flora of West Tropical Africa, revised edition by R. W. J. Keay (Vol. 1 1954–58) and F. N. Hepper (Vols. 2, 1963, & 3, 1968–72).

FZ. vol. as indicated: Flora Zambesiaca 1960 onwards.

Geiseler, E. F., 1807: Crotonis monographia.

Guillemin, J. B. A., S. Perrottet & A. Richard, 1830–33: Florae Senegambiae tentamen.

Hepper, F. N. (see FWTA.).

Hooker, W. J., 1836: Companion to the Botanical Magazine.
  —1849: Niger Flora.

Hornemann, J. W., 1807: Enumeratio Plantarum Horti botanici Hafniensis.
  —1809: Supplementum II. Enumerationis Plantarum Horti botanici Hafniensis.
  —1815: Hortus regis botanicus Hafniensis 2.
  —1819: De indole plantarum guineensium observationes.

Hutchinson, J. & J. M. Dalziel, 1927–36: Flora of West Tropical Africa, first edition.

Isert, P. E., 1788: Paul Erdmann Isert's ehemal. königl. dänisch. Oberarzte an den Besizzungen in Afrika, Reise nach Guinea und den Caribäischen Inseln in Columbien, in Briefen an seine Freunde beschrieben.
  —1789: Kurze Beschreibung und Abbildung einiger Vögel aus Guinea (Fortsetzung).
  —Schrift. Gesellsch. naturforsch. Freunde Berlin 9: 332–335, Pl. 9.

Jacquin, J. F., 1811–16: Eclogae plantarum rariorum 1 (to Pl. 100).
Jacquin, N. J., 1781–86: Icones plantarum rariorum 1 (to Pl. 200).
   —1786–93a: Icones plantarum rariorum 2 (to Pl. 454).
   —1786–93b: Icones plantarum rariorum 3.
   —1788: Observationes botanicae.—Collectanea 2: 260–374, Pl. 17, f. 1, 4,
      Pl. 18, f. 2, 3.
Junghans, J., 1961: Thonning's and Isert's collections from "Danish Guinea"
   (Ghana) in West Tropical Africa.—Botanisk Tidsskrift. 57: 310–355.
   —1962: as last 58: 82–122.
Jussieu, A. L., 1804: Mémoire sur le *Grewia*, genre de plantes de la famille des
   Tiliacées.—Ann. Mus. Nat. d'Hist. Naturelle 4: 82–93, Pl. 47–51.
Keay, R. W. J. (see FWTA.).
Murray, J. A., 1774: Systema vegetabilium, ed. 13.
Persoon, C. H., 1807: Synopsis plantarum 2.
Poiret, J. L. M., 1808: Encyclopédie méthodique.—Botanique 8.
   —1813: Encyclopédie méthodique.—Botanique, Suppl. 3.
Schubert, B., 1963: Comments on Junghans' list of Thonning's and Isert's collections.
   J. Arn. Arb. 44: 291–296.
Schumacher, F. C., 1827: Beskrivelse af guineiske Planter som ere fundne af danske
   Botanikera, isaer af Etatsraad Thonning.
   —1828: (The same paper in) Det Kongelige danske Videnskabernes Selskabs
      naturvidenskabelige og mathematiske Afhandlinger 3: 21–248.
   —1829: (continuation of last) 4: 1–236.
   —MS: (The manuscript of the same housed at Copenhagen Botanical Museum).
Vahl, M., 1791: Symbolae botanicae 2.
   —1794: Symbolae botanicae 3.
   —1796: Eclogae americanae 3.
   —1804: Enumeratio plantarum 1.
   —1805: Enumeratio plantarum 2.
   —1810: Beskrivelse over nye Planteslaegter, Skrivter af Naturhistorie-
      Selskabet 6: 84–128, Pl. 6.
   —MS: (The manuscript of Enumeratio plantarum; only two volumes of which
      were published).
Willdenow, K. L., 1792: Eine neue Pflanzengattung *Usteria* genannt.—Schrift.
   Gesellsch. naturforsch. Freunde Berlin 10: 51–56, Pl. 2.
   —1793 Duae plantae africanae. Usteri Delectus opusculorum botanicorum 2.
   —1794: Phytographia.
   —1796: Dialium guineense.—Römer's Archiv für die Botanik 1, 1: 30–32, Pl. 6.
   —1797a: Species plantarum 1, 1 (to p. 495).
   —1797b: Schradera et Rottlera.—Römer's Archiv für Botanik 1, 3: 131–135,
      Pl. 7.
   —1799: Species plantarum 2, (to p. 823).
   —1800a: Species plantarum 2, 2.
   —1800b: Species plantarum 3, 1 (to p. 850).
   —1802: Species plantarum 3, 2 (to p. 1470).
   —1803: Species plantarum 3, 3.
   —1805: Species plantarum 4, 1 (to p. 630).
   —1806a: Species plantarum 4, 2.
   —1806b: Hortus Berolinensis 1.
   —1809: Enumeratio plantarum horti regii botanici Berolinensis.
   —1810: Species plantarum 5, 1.

APPENDIX I

VERNACULAR (AND SOME EUROPEAN) NAMES	BOTANICAL EQUIVALENT
Abada	Uvaria chamae *P. Beauv.*
Abami-aulage-tio	Clausena anisata (*Willd.*) *Hook.f. ex Benth.*
Abami-tio	Clausena anisata (*Willd.*) *Hook.f. ex Benth.*
Abaumba (fruit)	Capparis erythrocarpos *Isert*
Abloge	Launaea taraxacifolia (*Willd.*) *Amin ex C. Jeffrey*
Abontaa	Ancylobotrys scandens (*Schum. et Thonn.*) *Pichon*
Adodomi	Spondias mombin *L.*
Aflaumbe (fruit)	Carissa edulis *Vahl*
Agingeli	Uvaria ovata (*Dunal*) *A. DC.* subsp. ovata
Ahablobae	Aloe picta *Thunb.*
Ahaemeté	Millettia irvinei *Hutch. et Dalz.*
Ajilebi	Ritchiea reflexa (*Thonning*) *Gilg et Benedict*
Akassi	Lonchocarpus cyanescens (*Schum. et Thonn.*) *Benth.*
Akokobessa (root)	Carissa edulis *Vahl*
Akoteno	Phoenix reclinata *Jacq.*
Alipoma-Kripei	Desmodium ramosissimum *G. Don*
Amagomi	Flacourtia flavescens *Willd.*
Amba-pang	Ipomoea pes-caprae (*L.*) *R. Br.* subsp. brasiliensis (*L.*) *Oostr.*
Amitjolobi	Phoenix reclinata *Jacq.*
Ammba-pang	Canavalia rosea (*Sw.*) *DC.*
Amotobi	Physalis angulata *L.*
Amuma	Eugenia coronata *Schum. et Thonn.*
Anmanum-ba	Cyphostemma cymosa (*Schum. et Thonn.*) *Descoings*
Anistree (European)	Clausena anisata (*Willd.*) *Hook.f. ex Benth.*
Apaataa	Ficus ovata *Vahl* (F. calyptrata)
Aquing	Voandzeia subterranea (*L.*) *Thouars*
Asablobae	Aloe picta *Thunb.*
Asianté-Kitteva	Abelmoschus moschatus *Medik.*
Asi-gremi	Grewia carpinifolia *Juss.*
Asogagaplae	Solanum anomalum *Thonning*
Assianthé boenner (Danish)	Arachis hypogaea *L.*
Atia-tjo	Blighia sapida *König*
Atropo	Solanum macrocarpon *L.*
Atropo-bah	Solanum dasyphyllum *Schum. et Thonn.*
Aumaadoati	Phyllanthus amarus *Schum. et Thonn.*
Aumbae	Diospyros tricolor (*Schum. et Thonn.*) *Hiern*
Azara-tjo	Oncoba spinosa *Forsk.*
Baabylaa-najrié	Nauclea latifolia *Sm.*
Badimanoplaa	Deinbollia pinnata (*Poir.*) *Schum. et Thonn.*
Bāsissa	Cassia occidentalis *L.*
Beseri	Dichrostachys cinerea (*L.*) *Wight et Arn.*
Blaa	Sansevieria liberica *Gér. et Labr.*

VERNACULAR (AND SOME EUROPEAN) NAMES	BOTANICAL EQUIVALENT
Blaabaa-fye	Asystasia calycina *Benth.*
Black tamarind (English)	Dialium guineense *Willd.*
Blafaa-Koae	Ocimum canum *Sims*
Blafo (Accra)	Tribulus terrestris *L.*
Blafo-atropo	Solanum melongena *L.*
Boj-tegi-tjo	Morinda lucida *Benth.*
Calevancus ('English')	Phaseolus adenanthus *G. F. W. Mey.*
Demi-tjo	Caesalpinia bonduc (*L.*) *Roxb.*
Dendrae	Solanum nigrum *L.*
Dojvié	Piper guineense *Thonning*
Eduásudoá	Tapinanthus bangwensis (*Engl. et K. Krause*) *Danser*
Engkatje	Arachis hypogaea *L.*
Enkafo	Sarcostemma viminale (*L.*) *R. Br.*
Fankvau	Solanum macrocarpon *L.*
Fjong	Vitex doniana *Sweet*
Fufuba	Waltheria indica *L.*
Gaegaenae	Pterocarpus santalinoides *L'Hérit. ex DC.*
Gubaa	Scaevola plumieri (*L.*) *Vahl*
Hah-tio	Fagara zanthoxyloides *Lam.*
Hallasjojo	Blumea aurita (*L. f.*) *DC.*
Imbebi	Sesuvium portulacastrum (*L.*) *L.*
Jangkumaetri	Jasminum dichotomum *Vahl*
Jankosno	Eulophia cristata (*Sw.*) *Steud.*
Jan-j'na	Momordica charantia *L.*
Jo	Phaseolus adenanthus *G. F. W. Mey.*
Joj-tjo	Dialium guineense *Willd.*
Jubbejubbe	Voandzeia subterranea (*L.*) *Thouars*
Jumo-saa	Cordia guineensis *Thonning*
Kaasaamae	Typha domingensis *Pers.*
Kah-ba	Pergularia daemia (*Forsk.*) *Chiov.*
Kahn-tjo	Dichrostachys cinerea (*L.*) *Wight et Arn.*
Keriro	Solenostemon monostachyus (*P. Beauv.*) *Briq.* subsp. monostachyus
Kina-tjo	Mitragyna inermis (*Willd.*) *Kuntze*
Kjaelae	Cola verticillata (*Thonning*) *Stapf ex A. Chev.*
Klovaké	Jacquemontia ovalifolia (*Vahl*) *Hall. f.*
Koae	Ocimum canum *Sims*
Koa-fyé	Solanum macrocarpon *L.*
Koina-fye	Corchorus aestuans *L.*
Koina-fye	Corchorus tridens *L.*
Koina-fye	Corchorus trilocularis *L.*
Koi-tio	Pavetta corymbosa (*DC.*) *F. N. Williams* var. corymbosa
Koklo-tjo	Afraegle paniculata (*Schum. et Thonn.*) *Engl.*
Kolly-tjo	Chassalia kolly (*Schum.*) *Hepper*
Koo-pang	Adenia lobata (*Jacq.*) *Engl.*
Kuppei kirei (Tamil)	Amaranthus viridis *L.*
Lablaku	Lonchocarpus sericeus (*Poir.*) *Kunth*
Laedjo-tjo	Albizia glaberrima (*Schum. et Thonn.*) *Benth.*
Lalaaba	Stachytarpheta angustifolia (*Mill.*) *Vahl*
Lavasa	Ehretia corymbosa *Bojer ex A. DC.*
Loeloa-pang	Ipomoea mauritiana *Jacq.*
Lomo-tjo	Securinega virosa (*Roxb. ex Willd.*) *Baill.*
Maajaa	Amaranthus viridis *L.*

VERNACULAR (AND SOME EUROPEAN) NAMES	BOTANICAL EQUIVALENT
Maej	Haemanthus multiflorus *Martyn*
Mah-tjo	Conocarpus erectus *L.*
Mem'lemeté	Pupalia lappacea (*L.*) *Juss.*
Me-tio	Ximenia americana *L.*
Mihaa	Haemanthus multiflorus *Martyn*
Molaque	Arachis hypogaea *L.*
Muteku	Avicennia africana *P. Beauv.*
Nàbaa-di	Newbouldia laevis (*P. Beauv.*) *Seemann ex Bureau*
Naba-tiolu	Erythrina senegalensis *DC.*
Naivié	Annona senegalensis *Pers.*
Nanni-adumatre	Cucumis melo *L.* var. agrestis *Naud.*
Nanni-jaa	Rhynchosia sublobata (*Schum. et Thonn.*) *Meikle*
Nanni-Kumi	Lantana camara *L.*
Nannis vandmelon (Danish)	Cucumis melo *L.* var. agrestis *Naud.*
Ninndu-tjo	Ficus ovata *Vahl*
Obosso-tjo	Premna quadrifolia *Schum. et Thonn.*
Odiboi	Eclipta prostrata (*L.*) *L.*
Odum-tjo	Ceiba pentandra (*L.*) *Gaertn.*
Ohoa-tjo	Drepanocarpus lunatus (*L. f.*) *G. F. W. Mey.*
Onjai-tjo	Ceiba pentandra (*L.*) *Gaertn.*
Osisiu	Spathodea campanulata *P. Beauv.*
Otiobibomo	Capparis brassii *DC.*
Ototrómi	Dichapetalum guineense (*DC.*) *Keay*
Otru	Sesamum alatum *Thonning*
Otru	Sesamum radiatum *Schum. et Thonn.*
Paettaeplae-bi	Gardenia ternifolia *Schum. et Thonn.*
Peber	Piper guineense *Thonning*
Pepraemese	Tetrapleura tetraptera (*Schum. et Thonn.*) *Taub.*
Petipeti	Capparis erythrocarpos *Isert*
Petipeti	Capparis tomentosa *Lam.*
Plem-tjo	Hippocratea africana (*Willd.*) *Loes. ex Engl.*
Plöm-tjo	Byrsocarpus coccineus *Schum. et Thonn.*
Quitto-bönner (Danish)	Phaseolus adenanthus *G. F. W. Mey*
Sablabó	Cardiospermum grandiflorum *Sw.*
Sablabó	Cardiospermum halicacabum L.
Sabullá	Allium guineense *Thonning*
Samangkama	Alternanthera pungens *Kunth*
Sebae	Solanum gilo *Raddi* var. gilo
Sia-pang	Kedrostis foetidissima (*Jacq.*) *Cogn.*
Sio-tahmi	Erythroxylum emarginatum *Thonning*
Sirsa-imum	Hibiscus surattensis *L.*
Sissa-imune	Hibiscus cannabinus *L.*
Sissa-koae	Orthosiphon suffrutescens (*Schum.*) *J. K. Morton*
Sissa-sussoa	Solanum anomalum *Thonning*
Sjaa-blaa	Scoparia dulcis *L.*
Sjadjo-tjo	Adansonia digitata *L.*
Sjo-tami	Byrsocarpus coccineus *Schum. et Thonn.*
Söd "vün palme" (Danish)	Phoenix reclinata *Jacq.*
Songu-tjo	Hyphaene guineensis *Schum. et Thonn.*
Sussoa	Solanum indicum *L.* subsp. distichum (*Thonning*) *Bitter*
Sylu	Ocimum gratissimum *L.*
Tadadua	Allophylus spicatus (*Poir.*) *Radlk.*

VERNACULAR (AND SOME EUROPEAN) NAMES	BOTANICAL EQUIVALENT
Taehn-tio	Elaeis guineensis *Jacq.*
Taeta-fye	Gynandropsis gynandra (*L.*) *Briq.*
Taetjoe-pang	Mucuna sloanei *Fawc. et Rendle*
Taetremande	Nymphaea lotus *L.*
Taetremande	Nymphaea maculata *Schum. et Thonn.*
Tahmi	Synsepalum dulcificum (*Schum. et Thonn.*) *Daniell*
Tah-tjo	Millettia thonningii (*Schumacher*) *Baker*
Tah-tjo	Vernonia colorata (*Willd.*) *Drake*
Tatadua	Allophylus africanus *P. Beauv.*
Tenjo-tjo	Elaeophorbia drupifera (*Thonning*) *Stapf*
Tjalala	Boerhavia diffusa *L.*
Toubé	Triumfetta rhomboidea *Jacq.*
Uaegte vün palme (Danish)	Phoenix reclinata *Jacq.*
Vild Kaschu (European)	Blighia sapida *König*
Vjye-tjo	Borassus aethiopum *Mart.*
Vualé-mi	Caesalpinia bonduc (*L.*) *Roxb.*
Vula-fyé	Ipomoea stolonifera (*Cyrill.*) *J. F. Gmel.*

APPENDIX II

Place-names mentioned by Isert & Thonning in Danish Guinea with modern equivalents.

Compiled in co-operation with J. B. Hall, Department of Botany, University of Ghana. All localities are in present-day Ghana, unless otherwise stated. See also Maps 1 and 2.

OLD DANISH NAME	MODERN EQUIVALENTS AND NOTES (All Ghana unless stated Dahomey)
Acra	= Accra 5 33N. 0 13W.
Ada, Adah	= Ada 5 47N. 0 38E.
Adampi	= Adangbe; tribe of people speaking language similar to the Ga spoken in Accra, living between Tema and Ada along the coast. Isert uses this name for Ningo, which is an Adangbe town.
Adami	= ? Adangbe (see last)
Adau	= ? Ada 5 47N. 0 38E.
Aflahu, Aflaha, Aflaku	= Aflao 6 07N. 1 11E.
Akim	= Akim (tribal area between Akwapim and Ashanti) off northern edge of map.
Akkra	= Accra 5 33N. 0 13W.
Akvambu	= Akwamu 6 17N. 0 05E.
Akvapim (see next)	
Aquapim	= Akwapim (more correctly, but less usually Akuapem); tribal area on the hills between Aburi and Koforidua, c. 6 00N. 0 12W.
Ashanti	= Ashanti (more correctly Asante) tribe and region.
Asohia	= ? Assoghé (Dahomey; 6 22N. 2 01E.).
Assiama, Assiamae, Assiame, Asiama	= either Ashaiman 5 42N. 0 02W. on the northern outskirts of Tema, or, more probably, Afiaman 5 42N. 0 17W. *Albizia glaberrima* was collected by Thonning 'between Asiama and Jadofa'; on Thonning's map the present Afiaman is shown as 'Assamang', and is across the end of the Akwapim Hills from 'Jadofa'; Ashaiman is shown as 'Assama', which is considerably further away from Jadofa.
Atacco, Atokke, Atocco	= Atoko 5 46N. 0 49E.
Augna Lake	= Keta Lagoon 5 54N. 0 56E.
Blegusso, Blekusu, Bligusso	= Berekuse 5 45N. 0 13W.
Christiansborg	= Osu (an eastern suburb of Accara) 5 33N. 0 11W.
Dadintam	= Adadientem 6 10N. 0 35W. This spelling is shown on Thonning's map; the place is much further from the coast than any of the other localities, lying near Kibi.
Dudna, Dudua, Dudum, Dudun	= Dodowa 5 53N. 0 07W.
Elmina	= Elmina 5 05N. 1 21W.
Fida	= Ouidah (Dahomey) 6 22N. 2 05E.
Fredensborg	= fort at Ningo

OLD DANISH NAME	MODERN EQUIVALENTS AND NOTES
Frederiksberg, Fredriksberg	=either the fort just inland from Osu and shown as ruined on Thonning's map; or the plantation near Abokobi 5 44N. o 13W., which was started in 1802 (after Isert's death), and visited by Thonning.
Frederikstad	=presumably fort just SW. of Dodowa, shown as ruined on Thonning's map. Isert called it Frederiks-nopel, later Frederikssted.
Ga, Gah	=Accra
Jadofa	=Oyarifa 5 44N. o 11W.
Kongesteen	=fort at Ada
Krat, Kratté	=perhaps the Danish word for thickets; see note under *Euphorbia lateriflora* by Keay in FWTA. ed. 2, 1 : 422.
Krepe, Krepé, Krepeh, Volta	=a place not traced, presumably near the Volta River
La, Labode, Labodei	=Labadi 5 33N. o 09W.
Lahtebierg	=Larte 5 56N. o 04W.
Lalve River	=Laiwi Lagoon 5 42N. o 04E.
Ningo	=Old Ningo 5 44N. o 13E. (also Nigo or Nugo on some modern maps)
Poisi	=Kpeshi Lagoon 5 33N. o 08W.
Popo	=Grand Popo (Dahomey) 6 17N. 1 50E.
Pottebra	=is mentioned by Isert as lying half way between Keta and Aflao, which would bring it close to the site of Adina 6 02N. 1 05E. It is quite possible that 'Pottebra' is based on error or misinformation, as it lay in the country of the people Isert's men were fighting against.
Prampram	=Prampram 5 42N. o 07E.
Prindsensteen, Prinzensteen	=fort at Keta
Quita, Quitta	=Keta 5 55N. o 59E.
Sakumo River	=Sakumo Lagoon. There are two lagoons of this name, one just to the west of Accra, 5 31N. o 19W. and the other just to the east, 5 37N. o 02W.
Sawi	=Savi (Dahomey) 6 25N. 2 06E.
Tessin	=Teshi 5 35N. o 06W.
Töffri	=Tefle 6 00N. o 34E.
Tubreku	=Not traced
Ursua, Ursue	=Osu (local name for Christiansborg).
Volta, Rio	=Volta River
Whyda, Whydah, Whydahr	=Ouidah (Dahomey) 6 22N. 2 05E.

MODERN NAME	EQUIVALENT NAME OR SPELLING GIVEN BY DANES
Abokobi	= ? Frederiksberg (see other list)
Accra	=Acra, Akkra, Ga, Gah
Ada	=Adah, ? Adau, Kongesteen (see other list)
Adadientem	=Dadintam
Adangbe	=Adampi, ? Adami
Adina (see other list under Pottebra)	
Aflao	=Aflaha, Aflahu, Aflaku
Afiaman	= ? Assiame, ? Assiama, ? Asiama
Akim (tribal area)	=Akim
Akwapim (tribal area)	=Akvapim, Aquapim
Ashaiman (see note on Assiame in other list)	

MODERN NAME	EQUIVALENT NAME OR SPELLING GIVEN BY DANES
Ashanti	=Ashanti
Assoghé	= ? Asohia
Atoko	=Atacco, Atokke, Atocco
Berekuso	=Blegusso, Blekusu, Bligusso
Dodowa	=Dudna, Dudua, Dudum, Dudun (see also Frederikstad in other list)
Elmina	=Elmina
Grand Popo	=Popo
Keta	=Quita, Quitta
Keta Lagoon	=Augna Lake
Kpeshi Lagoon	=Poisi
Labadi	=La, Labode, Labodei
Laiwi Lagoon	=Lalve River
Larte	=Lahtebierg
Ningo (fort)	=Fredensborg
Old Ningo	=Ningo
Ouidah	=Fida, Whyda, Whydah, Whydahr
Osu	=Christiansborg (see also Frederiksberg in other list), Ursu, Ursue
Oyarifa	=Jadofa
Prampram	=Prampram
Sakumo Lagoon	=Sakumo River
Savi	=Sawi
Tefle	=Töffri
Teshi	=Tessin
Volta River	=Rio Volta
Whydah	see Ouidah

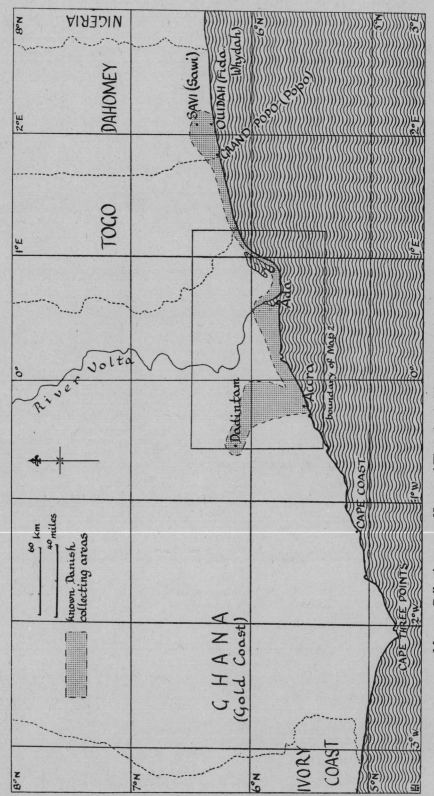

Map 1 Collecting areas of Isert and Thonning in West Africa showing localities in Danish Guinea

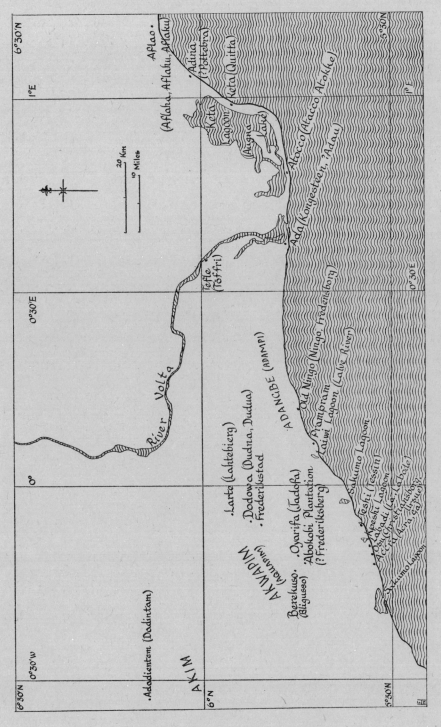

Map 2 Detail of collecting area in Ghana (formerly Danish Guinea)

APPENDIX III

LIST OF GENERA DESCRIBED AS NEW FROM THE ISERT AND THONNING
COLLECTIONS AND THEIR PRESENT STATUS

BALANOPHORACEAE
Thonningia *Vahl*—genus maintained
Type of genus: *T. sanguinea* Vahl
Type specimen: *Thonning* 94

CAESALPINIACEAE
Westia Vahl = **Berlinia** *Soland. ex Hook. f.*, nom. cons.
Type of genus: *W. grandiflora* Vahl = *Berlinia grandiflora* (Vahl) Hutch. et Dalz.
Type specimen: *Isert* s.n.

FLACOURTIACEAE
Lundia Schum. non DC. = **Oncoba** *Forsk.*
Type of genus = *L. monacantha* Schum. et Thonn. = *Oncoba spinosa* Forsk.
Type specimen: *Thonning* 296 (not traced)

PAPILIONACEAE
Sommerfeldia Schum. = **Drepanocarpus** *G. F. W. Mey.*
Type of genus: *S. obovata* Schum. et Thonn. = *Drepanocarpus lunatum* (Linn. f.)
    G. F. W. Mey.
Type specimen: *Thonning* 254

Rathkea Schum. = **Ormocarpum** *P. Beauv.*
Type of genus: *R. glabra* Schum. et Thonn. = *Ormocarpum sennoides* (Willd.) DC.
subsp. *hispidum* (Willd.) Brenan et J. Léonard

Plectrotropis Schum. et Thonn. = **Vigna** *Savi*
Type of genus: *P. angustifolia* Schum. et Thonn. = *Vigna vexillata* (L.) A. Rich.
Type specimen: *Thonning* 7

PASSIFLORACEAE
Buelowia Schum. = **Smeathmannia** *Soland. ex R. Br.*
Type of genus: *B. illustris* Schum. et Thonn. = *S. pubescens* Soland. ex R. Br.
Type specimen: *Thonning* 85 (not traced)

RUBIACEAE
Benzonia *Schum.*—see note on status, p. 105
Type of genus: *B. corymbosa* Schum.
Type specimen: *Thonning* s.n. (not traced)

Octodon Thonning = **Borreria** *G. F. W. Mey.*
Type of genus: *O. filifolium* Schum. et Thonn. = *Borreria filifolia* (Schum. et Thonn.)
    K. Schum.
Type specimen: *Thonning* 241

Cephalina Thonning = **Nauclea** *L.*
Type of genus: *C. esculenta* (Afzel. ex Sabine) Schum. et Thonn. = *N. latifolia* Sm.
Type specimen: *Afzelius* s.n.

Phallaria Schum. et Thonn. = **Canthium** *Lam.* and **Vangueriopsis** *Robyns ex Good*, pro parte.

Type species: *P. horizontalis* Schum. et Thonn. = *Canthium horizontale* Schum. et Thonn.) Hiern

Type specimen: *Thonning* s.n.

Type species: *P. spinosa* Schum. et Thonn. = *Vangueriopsis spinosa* (Schum. et Thonn.) Hepper

Type specimen: *Thonning* 287

SAPINDACEAE

Deinbollia *Schum.*—genus maintained

Type of genus: *D. pinnata* (Poir.) Schum. et Thonn.

Type specimen: *Thonning* 248

TURNERACEAE

Wormskioldia *Thonning*—genus maintained

Type of genus: *W. heterophylla* Schum. et Thonn. = *W. pilosa* (Willd.) Schweinf. ex Urb.

Type specimen: *Thonning* 297

URTICACEAE

Heynea *Schum.*—genus maintained

Type species: *H. ovalifolia* Schum. et Thonn. = *Laportea ovalifolia* (Schum. et Thonn.) Chew

Type specimen: *Thonning* 264

APPENDIX IV

1 Leonotis nepetifolia (*L.*) *Ait. f.* var. africana (*P. Beauv.*) *J. K. Morton* = *Phlomis pallida* Schum. et Thonn.
2 Merremia aegyptiaca (*L.*) *Urban* = *Convolvulus pentaphyllus* L.
3 Momordica charantia *L.* = *M. anthelmintica* Schum. et Thonn.
4 Phyllanthus amarus *Schum. et Thonn.*
5 Ipomoea coptica (*L.*) *Roth ex Roem. et Schult.* = *Convolvulus thonningii* Schumacher
6 Ipomoea ochracea (*Lindl.*) *G. Don* = *Convolvulus trichocalyx* Schum. et Thonn.
7 Vigna vexillata (*L.*) *A. Rich.* = *Plectrotropis angustifolia* Schum. et Thonn.
7 Plumbago zeylanica *L.* = *P. auriculata* Lam.
8 Cassytha filiformis *L.* = *Cassyta guineensis* Schum. et Thonn.
9 Curculigo pilosa (*Schum. et Thonn.*) *Engl.* = *Gethyllis pilosa* Schum. et Thonn.
10 Waltheria indica *L.* = *W. africana* Schumacher & *W. guineensis* Schumacher
11 Ipomoea pes-caprae (*L.*) *R. Br.* subsp. brasiliensis (*L.*) *Oostr.* = *Convolvulus rotundifolius* Schum. et Thonn.
12 Cassia occidentalis *L.* = *C. planisiliqua* L.
13 Urena lobata *L.* = *U. diversifolia* Schum. et Thonn.
14 Byrsocarpus coccineus *Schum. et Thonn.* = *B. puniceus* Schum. et Thonn.
15 Millettia thonningii (*Schumacher*) *Baker* = *Robinia thonningii* Schumacher
16 Mitragyna inermis (*Willd.*) *Kuntze* = *Nauclea africana* Willd.
16 (partly in error) Conocarpus erectus *L.*
17 Boerhavia diffusa *L.*
18 Hibiscus vitifolius *L.* = *H. strigosus* Schum. et Thonn.
19 Byrsocarpus coccineus *Schum. et Thonn.*
20 Indigofera tinctoria *L.* = *I. ornithopodioides* Schumacher
21 Ipomoea mauritiana *Jacq.* = *Convolvulus paniculatus* L.
22 Scaevola plumieri (*L.*) *Vahl* = *S. lobelia* Murr.
23 Ocimum gratissimum *L.* = *O. guineense* Schum. et Thonn.
24 Grewia carpinifolia *Juss.*
25 Triumfetta rhomboidea *Jacq.* = *T. mollis* Schum. et Thonn.
26 Crateva adansonii *DC.* = *Crataeva guineensis* Schum. et Thonn.
27 Phyllanthus maderaspatensis *L.* = *P. thonningii* Schumacher
28 Ruspolia hypocrateriformis (*Vahl*) *Milne-Redh.* = *Justicia hypocrateriformis* Vahl
29 Clausena anisata (*Willd.*) *Hook. f. ex Benth.* = *Amyris anisata* Willd.
30 Tribulus terrestris *L.* = *T. humifusus* Schum. et Thonn.
31 Ocimum canum *Sims* = *O. lanceolatum* Schum. et Thonn.
32 Euphorbia glaucophylla *Poir.* = *E. trinervia* Schum. et Thonn.
33 Securinega virosa (*Roxb. ex Willd.*) *Baill.* = *Phyllanthus angulatus* Schum. et Thonn.
34 Phyllanthus pentandrus *Schum. et Thonn.*
35 Indigofera hirsuta *L.* var. hirsuta = *I. ferruginea* Schum. et Thonn.
36 Indigofera dendroides *Jacq.*
37 Phyllanthus sublanatus *Schum. et Thonn.*
38 Abutilon guineense (*Schum. et Thonn.*) *Bak. f.* = *Sida guineensis* Schum. et Thonn.
39 Amaryllis trigona *Thonning*
40 Crotalaria pallida *Ait.* var. pallida = *C. striata* DC.
41 Indigofera tetrasperma *Vahl ex Pers.*

42   Indigofera spicata *Forsk.* = *I. hendecaphylla* Jacq.

43   Sesbania sericea (*Willd.*) *Link* = *Emerus pubescens* (DC.) Schum. et Thonn.

44   Uvaria chamae *P. Beauv.* = *U. cylindrica* Schum. et Thonn.

45   Indigofera subulata *Vahl ex Poir.* var. subulata = *I. thonningii* Schum. et Thonn.

46   Fagara zanthoxyloides *Lam.* = *Zanthoxylum polygamum* Schum. et Thonn.

47   Ormocarpum sennoides (*Willd.*) *DC.* subsp. hispidum (*Willd.*) *Brenan et J. Léonard* = *Rathkea glabra* Schum. et Thonn.

48   Baphia nitida *Lodd.* = *Podalyria haematoxylon* Thonning

49   Cassia mimosoides *L.* = *C. geminata* Vahl ex Schum. et Thonn.

50   Ophioglossum costatum *R. Br.* = *O. fibrosum* Schumacher

51   Phyllanthus reticulatus *Poir.* var. glaber *Müll. Arg.* = *P. polyspermus* Schum. et Thonn.

52   Melanthera scandens (*Schum. et Thonn.*) *Roberty* = *Buphthalmum scandens* Schum. et Thonn.

53   Eugenia coronata *Schum. et Thonn.*

54   Adansonia digitata *L.*

54   Chrysobalanus orbicularis *Schumacher*

55   Flacourtia flavescens *Willd.* = *F. edulis* Schum. et Thonn.

56   Heliotropium strigosum *Willd.*

57   Allophylus spicatus (*Poir.*) *Radk.* = *Ornithrophe magica* Schum. et Thonn.

58   Habenaria filicornis (*Thonning*) *Lindl.* = *Orchis filicornis* Thonning

59   Melia azedarach *L.* = *M. angustifolia* Schum. et Thonn.

60   Caesalpinia pulcherrima (*L.*) *Sw.*

61   Indigofera nigricans *Vahl ex Pers.* = *I. elegans* Schum. et Thonn.

62   Jacquemontia ovalifolia (*Vahl*) *Hall. f.* = *Convolvulus coeruleus* Schum. et Thonn.

63   Cassia absus *L.* = *C. viscosa* Schum. et Thonn.

64   Canavalia rosea (*Sw.*) *DC.* = *Dolichos obovatus* Schum. et Thonn.

65   Mimosa pigra *L.* = *M. procumbens* Schum. et Thonn.

66   Trianthema portulacastrum *L.* = *T. flexuosa* Schum. et Thonn.

67   Gynandropsis gynandra (*L.*) *Briq.* = *Cleome acuta* Schum. et Thonn.

68   Glinus oppositifolius (*L.*) *A. DC.* = *Pharnaceum mollugo* L.

69   Heliotropium indicum *L.* = *H. africanum* Schum. et Thonn.

70   Polygala arenaria *Willd.*

71   Polygala guineensis *Willd.*

72   Ouratea flava (*Schum. et Thonn.*) *Hutch. et Dalz.* = *Gomphia flava* Schum. et Thonn.

73   Hybanthus thesiifolius (*Juss. ex Poir.*) *Hutch. et Dalz.* = *Viola guineensis* Schum. et Thonn.

74   Commelina erecta *L.* = *C. umbellata* Thonning

75   Murdannia simplex (*Vahl*) *Brenan* = *Commelina simplex* Vahl

76   Jasminum dichotomum *Vahl*

77   Lonchocarpus cyanescens (*Schum. et Thonn.*) *Benth.* = *Robinia cyanescens* Schum. et Thonn.

78   Endostemon tereticaulis (*Poir.*) *M. Ashby* = *Ocymum thonningii* Schumacher

79   Carissa edulis *Vahl* = *C. dulcis* Schum. et Thonn.

80   Uvaria ovata (*Dunal*) *A. DC.* subsp. ovata = *U. cordata* Schum. et Thonn.

81   Indigofera pulchra *Willd.*

82   Ludwigia octovalvis (*Jacq.*) *Raven* subsp. brevisepala (*Brenan*) *Raven* = *Jussieua linearis* Willd.

83   Spondias mombin *L.* = *S. aurantiaca* Schum. et Thonn.

84   Nauclea latifolia *Sm.* = *Cephalina esculenta* Schum. et Thonn.

85   Momordica foetida *Schumacher*

85   Smeathmannia pubescens *Soland. ex R. Br.* = *Buelowia illustris* Schum. et Thonn.

86   Indigofera macrophylla *Schumacher*

87 Merremia tridentata (*L.*) *Hall. f.* subsp. angustifolia (*Jacq.*) *Oostr.* =*Convolvulus filicaulis* Vahl

88 Hibiscus micranthus *L. f.* =*H. versicolor* Schum. et Thonn.

89 Ehretia corymbosa *Bojer ex A. DC.* =*E. cymosa* Thonning

90 Tetrapleura tetraptera (*Schum. et Thonn.*) *Taub.* = *Adenanthera tetraptera* Schum. et Thonn.

91 Ximenia americana *L.*

92 Solenostemon monostachyus (*P. Beauv.*) *Briq.* subsp. monostachyus = *S. ocymoides* Schum. et Thonn.

93 Mussaenda erythrophylla *Schum. et Thonn.*

94 Thonningia sanguinea *Vahl*

95 Spathodea campanulata *P. Beauv.* =*Bignonia tulipifera* Thonning

96 Griffonia simplicifolia (*Vahl ex DC.*) *Baill.* =*Schotia simplicifolia* Vahl ex DC.

97 Boerhavia diffusa *L.* =*B. adscendens* Willd.

98 Avicennia germinans (*L.*) *L.*

99 Capparis erythrocarpos *Isert*

100 Ritchiea reflexa (*Thonning*) *Gilg et Benedict* =*Capparis reflexa* Thonning

101 Phoenix reclinata *Jacq.* =*P. spinosa* Schum. et Thonn.

102 Portulaca foliosa *Ker-Gawl.* =*P. prolifera* Schum. et Thonn.

103 Aponogeton subconjugatus *Schum. et Thonn.*

104 Erythrina senegalensis *DC.* = *E. latifolia* Schum. et Thonn.

105 Dialium guineense *Willd.* =*Codarium nitidum* Sol. ex Vahl

106 Iphigenia ledermannii *Engl. et K. Krause* =*Helonias guineensis* Thonning

107 Launaea taraxacifolia (*Willd.*) *Amin ex C. Jeffrey*

108 Millettia irvinei *Hutch. et Dalz.* =*Robinia multiflora* Schum. et Thonn.

109 Physalis angulata *L.*

110 Jatropha curcas *L.*

111 Scoparia dulcis *L.*

112 Abelmoschus moschatus *Medik.* =*Hibiscus abelmoschus* L.

113 Sophora occidentalis *L.* =*S. nitens* Schum. et Thonn.

114 Schwenckia americana *L.* = *S. guineensis* Schum. et Thonn.

115 Cleome rutidosperma *DC.* =*C. ciliata* Schum. et Thonn.

116 Croton lobatus *L.* =*C. trilobatum* Forsk.

117 Solanum macrocarpon *L.* =*S. atropo* Schum. et Thonn.

118 Aspilia helianthoides (*Schum. et Thonn.*) *Oliv. et Hiern* =*Coronocarpus helianthoides* Schum. et Thonn.

119 Sida cordifolia *L.* =*S. decagyna* Schum. et Thonn.

120 Sida linifolia *Juss. ex Cav.* =*S. linearifolia* Thonning

121 Wissadula amplissima *R. E. Fries* var. rostrata (*Schum. et Thonn.*) *R. E. Fries* = *Sida rostrata* Schum. et Thonn.

122 Cardiospermum halicacabum *L.* =*C. glabrum* Schum. et Thonn.

123 Newbouldia laevis (*P. Beauv.*) *Seemann ex Bureau* =*Bignonia glandulosa* Schum. et Thonn.

124 Arachis hypogaea *L.*

125 Lantana camara *L.* =*L. antidotalis* Schum. et Thonn.

126 Tapinanthus bangwensis (*Engl. et K. Krause*) *Danser* =*Loranthus thonningii* Schumacher

127 Indigofera aspera *Perr. ex DC.* =*I. tenella* Schumacher

128 Annona senegalensis *Pers.* subsp. senegalensis =*A. arenaria* Thonning

129 Uraria picta (*Jacq.*) *DC.* =*Hedysarum pictum* Jacq.

130 Hibiscus cannabinus *L.* =*H. congener* Schum. et Thonn.

131 Vernonia colorata (*Willd.*) *Drake* =*Chrysocoma amara* Schum. et Thonn.

132 Sesuvium portulacastrum (*L.*) *L.* =*S. brevifolium* Schum. et Thonn.

133 Dichrostachys cinerea (*L.*) *Wight et Arn.* = *Mimosa bicolor* Schum. et Thonn.

134 Solanum nigrum *L.* =*S. nodiflorum* Jacq.

135   Solanum anomalum *Thonning*
136   Combretum racemosum *P. Beauv.* = *C. corymbosum* Schumacher
137   Corchorus tridens *L.* = *C. angustifolius* Schum. et Thonn.
138   Corchorus trilocularis *L.* = *C. muricatus* Schum. et Thonn.
139   Corchorus aestuans *L.* = *C. polygonus* Schum. et Thonn.
140   Gardenia ternifolia *Schum. et Thonn.*
141   Solanum melongena *L.* = *S. edule* Schum. et Thonn.
142   Alternanthera sessilis (*L.*) *DC.* = *Illecebrum sessile* (*L.*) L.
143   Solanum gilo *Raddi* var. gilo = *S. geminifolium* Thonning
144   Solanum dasyphyllum *Schum. et Thonn.*
145   Phaseolus adenanthus *G. F. W. Mey.* = *Dolichos oleraceus* Schum. et Thonn.
146   Mucuna sloanei *Fawc. et Rendle* = *Stizolobium urens* (L.) Pers.
147   Aeschynomene indica *L.* = *A. quadrata* Schum. et Thonn.
148   Pergularia daemia (*Forsk.*) *Chiov* = *Asclepias convolvulacea* Willd.
149   Nymphaea guineensis *Schum. et Thonn.*
150   Blighia sapida *König* = *Cupania edulis* Schum. et Thonn.
151   Cucumis melo *L.* var. agrestis *Naud.* = *C.arenarius* Schum. et Thonn.
152   Limnophyton obtusifolium (*L.*) *Miq.* = *Alisma sagittifolia* Willd.
153   Zehneria hallii *C. Jeffrey* = *Bryonia deltoidea* Schum. et Thonn.
154   Calotropis procera (*Ait.*) *Ait. f.* = *Asclepias procera* Ait.
155   Cardiospermum grandiflorum *Sw.* = *C. hirsutum* Willd.
156   Paullinia pinnata *L.* = *P. uvata* Schum. et Thonn.
157   Argemone mexicana *L.*
158   Combretum smeathmannii *G. Don* = *C. mucronatum* Schum. et Thonn.
159   Phyllanthus capillaris *Schum. et Thonn.*
160   Ceiba pentandra (*L.*) *Gaertn.* = *Bombax guineense* Thonning
161   Premna quadrifolia *Schum. et Thonn.*
162   Evolvulus alsinoides (*L.*) *L.* = *E. azureus* Vahl et Thonn.
163   Pupalia lappacea (*L.*) *Juss.* = *Achyranthes mollis* Thonn.
164   Conocarpus erectus *L.* = *C. pubescens* Schum. et Thonn.
165   Philoxerus vermicularis (*L.*) *P. Beauv.* = *Gomphrena cylindrica* Schum. et Thonn.
166   Lonchocarpus sericeus (*Poir.*) *Kunth* = *Robinia argentiflora* Schum. et Thonn.
167   Ceiba pentandra (*L.*) *Gaertn.* = *Bombax pentandrum* L.
168   Synsepalum dulcificum (*Schum. et Thonn.*) *Daniell* = *Bumelia dulcifica* Schum. et Thonn.
169   Amaryllis nivea *Thonning*
170   Mussaenda elegans *Schum. et Thonn.*
171   Euadenia trifoliolata (*Vahl et Thonn.*) *Oliv.* = *Stroemia trifoliata* Vahl et Thonn.
172   Annona glauca *Schum. et Thonn.*
173   Acridocarpus alternifolius (*Schum. et Thonn.*) *Niedenzu* = *Malpighia alternifolia* Schum. et Thonn.
174   Ophrestia hedysaroides (*Willd.*) *Verdc.* = *Glycine hedysaroides* Willd.
175   Nymphaea maculata *Schum. et Thonn.*
176   Gisekia pharnaceoides *L.* = *G. linearifolia* Schum. et Thonn.
177   Oldenlandia corymbosa *L.* = *Hedyotis longifolia* Schum. et Thonn.
178   Kohautia virgata (*Willd.*) *Bremek.* = *Hedyotis virgata* Willd.
179   Cyphostemma cymosa (*Schum. et Thonn.*) *Descoings* = *Cissus cymosa* Schum. et Thonn.
180   Celosia trigyna *L.* = *C. laxa* Schum. et Thonn.
181   Ipomoea sepiaria *Roxb.* = *Convolvulus diversifolius* Schum. et Thonn.
182   Crotalaria glauca *Willd.*
183   Nymphaea lotus *L.* = *N. dentata* Schum. et Thonn.
184   Blepharis maderaspatensis (*L.*) *Heyne ex Roth* subsp. rubiifolia (*Schum. et Thonn.*) *Napper* = *B. rubiaefolia* Schum. et Thonn.

185 Hygrophila auriculata (*Schum. et Thonn.*) *Heine* = *Barleria auriculata* Schum. et Thonn.

186 Pandiaka angustifolia (*Vahl*) *Hepper* = *Gomphrena angustifolia* Vahl

187 Desmodium triflorum (*L.*) *DC.* = *Hedysarum granulatum* Schum. et Thonn.

188 Alysicarpus rugosus (*Willd.*) *DC.* = *Hedysarum rugosum* Willd.

189 Crotalaria goreensis *Guill. et Perr.* = *C. falcata* Schumacher

190 Jacquemontia tamnifolia (*L.*) *Griseb.* = *Convolvulus guineensis* Schumacher

191 Borreria scabra (*Schum. et Thonn.*) *K. Schum.* = *Diodia scabra* Schum. et Thonn.

192 Mollugo nudicaulis *Lam.* = *Pharnaceum spathulatum* Sw.

193 Alternanthera pungens *Kunth* = *Illecebrum obliquum* Schum. et Thonn.

194 Rhynchosia minima (*L.*) *DC.* = *Glycine rhombea* Schum. et Thonn.

195 Calliandra portoricensis (*Jacq.*) *Benth.* = *Mimosa guineensis* Schum. et Thonn.

196 Hoslundia opposita *Vahl et Thonn.*

197 Hibiscus surattensis *L.*

198 Hydrolea glabra *Schum. et Thonn.*

199 Solanum indicum *L.* subsp. distichum (*Schum. et Thonn.*) *Bitter* = *S. distichum* Schum. et Thonn.

200 Pergularia daemia (*Forsk.*) *Chiov.* = *Asclepias muricata* Schum. et Thonn.

201 Desmodium gangeticum (*L.*) *DC.* = *Hedysarum lanceolatum* Schum. et Thonn.

202 Desmodium velutinum (*Willd.*) *DC.* = *Hedysarum deltoideum* Schum. et Thonn.

203 Desmodium ramosissimum *G. Don* = *Hedysarum fruticulosum* Schum. et Thonn.

204 Indigofera paniculata *Vahl ex Pers.* = *I. procera* Schum. et Thonn.

205 Aerva lanata (*L.*) *Juss. ex Schult.* = *Illecebrum lanatum* (*L.*) Murr.

206 Aneilema umbrosum (*Vahl*) *Kunth* = *Commelina umbrosa* Vahl

207 Bidens pilosa *L.* = *B. abortiva* Schum. et Thonn.

208 Hewittia sublobata (*L. f.*) *Kuntze* = *Convolvulus involucratus* Willd.

209 Ipomoea involucrata *P. Beauv.* = *Convolvulus perfoliatus* Schum. et Thonn.

210 Oldenlandia lancifolia (*Schumacher*) *DC.* = *Hedyotis lancifolia* Schum.

211 Pentodon pentandrus (*Schum. et Thonn.*) *Vatke* = *Hedyotis pentandra* Schum. et Thonn.

212 Diodia sarmentosa *Sw.* = *D. pilosa* Schum. et Thonn.

213 Vigna vexillata (*L.*) *A. Rich.* = *Plectrotropis hirsuta* Schum. et Thonn.

214 Justicia flava (*Forsk.*) *Vahl* = *J. plicata* Vahl et Thonn.

215 Alternanthera pungens *Kunth* = *A. thonningii* Schumacher

216 Cyathula achyranthoides (*Kunth*) *Moq.* = *Achyranthes geminata* Thonning

217 Celosia argentea *L.* = *C. splendens* Schum. et Thonn.

218 Trema orientalis (*L.*) *Bl.* = *Celtis guineensis* Schum. et Thonn.

219 Hybanthus enneaspermus (*L.*) *F. v. Müll* = *Viola lanceifolia* Thonning

220 Tristemma incompletum *R. Br.* = *Melastoma sessilis* Schum. et Thonn.

221 Canavalia virosa (*Roxb.*) *Wight et Arn.* = *Dolichos ovalifolius* Schum. et Thonn.

222 Tragia monadelpha *Schum. et Thonn.*

223 Platostoma africanum *P. Beauv.* = *Ocymum sylvaticum* Thonning

224 Hyptis lanceolata *Poir.* = *H. lanceifolia* Thonning

225 Clerodendrum capitatum (*Willd.*) *Schum. et Thonn.* var. capitatum = *Volkameria capitata* Willd.

226 Artanema longifolium (*L.*) *Vatke* = *Achimenes sesamoides* Vahl

227 Spilanthes filicaulis (*Schum. et Thonn.*) *C. D. Adams* = *Eclipta filicaulis* Schum. et Thonn.

228 Erigeron spathulatum *Schum. et Thonn.*

229 Allophylus africanus *P. Beauv.* = *Ornithrophe tristachyos* Schum. et Thonn.

230 Ipomoea stolonifera (*Cyrill.*) *J. F. Gmel.* = *Convolvulus incurvus* Schum. et Thonn.

231 Stylosanthes erecta *P. Beauv.* = *S. guineensis* Schum. et Thonn.

232 Caesalpinia bonduc (*L.*) *Roxb.* = *Guilandina bonducella* L.

233 Capparis tomentosa *Lam.*

234 Capparis brassii *DC.* = *C. thonningii* Schumacher

235 Dodonaea viscosa *Jacq.* = *D. repanda* Schum. et Thonn.
236 Vahlia dichotoma (*Murr.*) *Kuntze* = *Oldenlandia pentandra* sensu Schum.
237 Euphorbia purpurascens *Schum. et Thonn.*
238 Sansevieria liberica *Gér. et Labr.* = *S. guineensis* sensu Willd.
239 Acacia nilotica (*L.*) *Willd. ex Del.* subsp. adstringens (*Schum. et Thonn.*) *Roberty et Brenan* = *Mimosa adstringens* Schum. et Thonn.
240 Enydra fluctuans *Lour.* = *Wahlenbergia globularis* Schum. et Thonn.
241 Borreria filifolia (*Schum. et Thonn.*) *K. Schum.* = *Octodon filifolium* Schum. et Thonn.
242 Mitracarpus villosus (*Sw.*) *DC.* = *Staurospermum verticillatum* Schum. et Thonn.
243 Polycarpaea stellata (*Willd.*) *DC.* = *Mollia stellata* (Willd.) Willd.
244 Vitex doniana *Sweet* = *V. cuneata* Thonning
245 Hippocratea africana (*Willd.*) *Loes. ex Engl.* = *Tonsella africana* Willd.
246 Ficus ovata *Vahl*
247 Rhynchosia sublobata (*Schum. et Thonn.*) *Meikle* = *Glycine sublobata* Schum. et Thonn.
248 Deinbollia pinnata (*Poir.*) *Schum. et Thonn.*
249 Coldenia procumbens *L.*
250 Adenia lobata (*Jacq.*) *Engl.* = *Modekka diversifolius* Schum. et Thonn.
251 Leptadenia hastata (*Pers.*) *Decne.* = *Cynanchum lancifolium* Schum. et Thonn.
252 Diospyros tricolor (*Schum. et Thonn.*) *Hiern* = *Noltia tricolor* Schum. et Thonn.
253 Ipomoea cairica (*L.*) *Sweet* = *Convolvulus cairicus* L.
254 Drepanocarpus lunatus (*L. f.*) *G. F. W. Mey.* = *Sommerfeldtia obovata* Schum. et Thonn.
255 Diodia serrulata (*P. Beauv.*) *G. Tayl.* = *D. maritima* Thonning
256 Eulophia cristata (*Sw.*) *Steud.* = *Limodorum articulatum* Schum. et Thonn.
257 Haemanthus cruentatus *Schum. et Thonn.*
258 Borassus aethiopium *Mart.* = *B. flabelliformis* Murr.
259 Hilleria latifolia (*Lam.*) *H. Walt.* = *Rivina apetala* Schum. et Thonn.
260 Sida acuta *Burm. f.* = *S. rugosa* Thonning
261 Eulophia gracilis *Lindl.* = *Limodorum ciliatum* Schum. et Thonn.
262 Motandra guineensis (*Thonning*) *A. DC.* = *Echites guineensis* Thonning
263 Trichilia monadelpha (*Thonning*) *De Wilde* = *Limonia monadelpha* Thonning
264 Laportea ovalifolia (*Schum. et Thonn.*) *Chew* = *Haynea ovalifolia* Schum. et Thonn.
264 Urera mannii (*Wedd.*) *Benth. et Hook. f. ex Rendle*
265 Vitex ferruginea *Schum. et Thonn.*
266 Elaeophorbia drupifera (*Thonning*) *Stapf* = *Euphorbia drupifera* Thonning
267 Amaranthus viridis *L.* = *A. polystachyus* Willd.
268 Cordia guineensis *Thonning*
269 Conyza aegyptiaca (*L.*) *Ait.* = *Erigeron exstipulatum* Schum. et Thonn.
270 Alafia scandens (*Thonning*) *De Wild.* = *Nerium scandens* Thonning
271 Pavetta corymbosa (*DC.*) *F. N. Williams* var. corymbosa = *Ixora nitida* Schum. et Thonn.
272 Albizia glaberrima (*Schum. et Thonn.*) *Benth.* = *Mimosa glaberrima* Schum. et Thonn.
273 Albizia adianthifolia (*Schumacher*) *W. F. Wight* = *Mimosa adianthifolia* Schumacher
274 Acacia pentagona (*Schum. et Thonn.*) *Hook. f.* = *Mimosa pentagona* Schum. et Thonn.
275 Acalypha ciliata *Forsk.* = *A. fimbriata* Schum. et Thonn.
276 Sesamum radiatum *Schum. et Thonn.*
277 Sesamum alatum *Thonning*
278 Luffa aegyptica *Mill.* = *L. scabra* Schum. et Thonn.
279 Urginea ensifolia (*Thonning*) *Hepper* = *Ornithogalum ensifolium* Thonning
280 Leea guineensis *G. Don* = *L. sambucina* sensu Schum.

281 Ancylobotrys scandens (*Schum. et Thonn.*) *Pichon* =*Strychnos scandens* Schum. et Thonn.

282 Piper guineense *Thonning*

283 Cola verticillata (*Thonning*) *Stapf ex A. Chev.* =*Sterculia verticillata* Thonning

284 Striga linearifolia (*Schum. et Thonn.*) *Hepper* =*Buchnera linearifolia* Schum. et Thonn.

285 Dissotis rotundifolia (*Sm.*) *Triana* var prostrata (*Thonning*) *Jac.-Fél.* = *Melastoma prostrata* Thonning

286 Oxyanthus racemosus (*Schum. et Thonn.*) *Keay* =*Ucriana racemosa* Schum. et Thonn.

287 Vanguieriopsis spinosa (*Schum. et Thonn.*) *Hepper* =*Phallaria spinosa* Schum. et Thonn.

288 Orthosiphon suffrutescens (*Schumacher*) *J. K. Morton* =*Ocymum suffrutescens* Schumacher

289 Dichapetalum guineense (*DC.*) *Keay* =*Rhamnus paniculatus* Thonning

290 Erythroxylum emarginatum *Thonning*

291 Afraegle paniculata (*Schum. et Thonn.*) *Engl.* =*Citrus paniculata* Schum. et Thonn.

292 Cissus quadrangularis *L.* =*C. triandra* Schum. et Thonn.

293 Tacca leontopetaloides (*L.*) *Kuntze* = *T. involucrata* Schum. et Thonn.

294 Cassia obtusifolia *L.*

295 Ocimum canum *Sims* = *O. hispidulum* Schum. et Thonn.

296 Oncoba spinosa *Forsk.* =*Lundia monacantha* Schum et Thonn.

297 Wormskioldia pilosa (*Willd.*) *Schweinf. ex Urb.* = *W. heterophylla* Schum. et Thonn.

298 Euphoria lateriflora *Schum. et Thonn.*

299 Cremaspora triflora (*Thonning*) *K. Schum.* =*Psychotria triflora* Thonning

300 Canthium multiflorum (*Schum. et Thonn.*) *Hiern* =*Psychotria multiflora* Thonning

301 Cissus quadrangularis *L.* =*C. bifida* Schum. et Thonn.

301 Ctenitis protensa (*Afzel. ex Sw.*) *Ching* =*Polypodium pubescens* Schumacher

302 Ctenitis cirrhosa (*Schum.*) *Ching* =*Aspidium cirrhosum* Schumacher

302 Hibiscus owariensis *P. Beauv.* =*H. triumfettaefolius* Thonning

303 Sarcostemma viminale (*L.*) *R. Br.* =*Asclepias nuda* Schum. et Thonn.

304 Tectaria angelicifolia (*Schumacher*) *Copel.* =*Polypodium angelicaefolium* Schumacher

304 Sesbania pachycarpa *DC. emend. Guill. et Perr.* =*Emerus aculeata* (Willd.) Hornem.

305 Pteris acanthoneura *Alston* =*P. spinulifera* Schumacher

305 Datura metel *L.* =*D. fastuosa* L.

306 Stachytarpheta angustifolia (*Mill.*) *Vahl* =*S. indica* sensu Schumacher

307 Doryopteris kirkii (*Hook.*) *Alston* =*Adiantum palmatum* Schumacher

307 Indigofera pilosa *Poir.* =*I. guineensis* Schum. et Thonn.

308 Gymnema sylvestre (*Retz.*) *Schultes* =*Cynanchum subvolubile* Schum. et Thonn.

308 Arthropteris palisotii *Desv.* =*Adiantum sublobatum* Schumacher

309 Basilicum polystachyon (*L.*) *Moench.* = *Ocymum dimidiatum* Schum. et Thonn.

310 Leucas martinicensis (*Jacq.*) *Ait. f.* =*Phlomis mollis* Schum. et Thonn.

311 Vernonia cinerea (*L.*) *Less.* =*Chrysocoma violacea* Schum. et Thonn.

312 Allium guineense *Thonning*

313 Crassocephalum rubens (*Juss. ex Jacq.*) *S. Moore* =*Cacalia uniflora* Schum. et Thonn.

314 Vacant

315 Quisqualis indica *L.* =*Q. obovata* Schum. et Thonn.

316 Alchornea cordifolia (*Schum. et Thonn.*) *Müll. Arg.* =*Schousboea cordifolia* Schum. et Thonn.

317 Eclipta alba (*L.*) *Hassk.* =*E. punctata* L.

318 Zehneria capillacea (*Schum. et Thonn.*) *C. Jeffrey* =*Bryonia capillacea* Schum. et Thonn.

319 Dalbergia ecastaphyllum (*L.*) *Taub.* =*Ecastaphyllum brownei* Pers.

320   Hyptis pectinata (*L.*) *Poit.* = *Bystropogon coarctatus* Schum. et Thonn.
321   Phragmanthera incana (*Schumacher*) *Balle* = *Loranthus incanus* Schum. et Thonn.
322   Psychotria calva *Hiern* = *P. umbellata* Thonning
323   Maytenus undata (*Thunb.*) *Blakelock* = *Celastrus lancifolius* Thonning
324   Ficus lutea *Vahl*
325   Ficus thonningii *Blume* = *F. microcarpa* Vahl
326   Ficus umbellata *Vahl*
327   Gossypium barbadense *L.* = *G. punctatum* Schum. et Thonn.
328   Gossypium herbaceum *L.* var. acerifolium (*Guill. et Perr.*) *A. Chev.* = *G. prostratum*
          Schum. et Thonn.
329   Ficus ovata *Vahl* = *F. calyptrata* Vahl
330   Asystasia quaterna (*Thonning*) *Nees* = *Ruellia quaterna* Thonning
331   Utricularia inflexa *Forsk.* var. inflexa = *U. thonningii* Schumacher
332   Hyphaene thebaica (*L.*) *Mart.* = *H. guineensis* Schum. et Thonn.
333   Elaeis guineensis *Jacq.*
334   Bacopa crenata (*P. Beauv.*) *Hepper* = *Erinus africanus* L.
335   Striga aspera (*Willd.*) *Benth.* = *Buchnera aspera* (Willd.) Schumacher
336   Micrargeria filiformis (*Schum. et Thonn.*) *Hutch. et Dalz.* = *Gerardia filiformis*
          Schum. et Thonn.
337   Kedrostis foetidissima (*Jacq.*) *Cogn.* = *Bryonia foetidissima* Schum. et Thonn.
338   Enteropogon macrostachyus (*Hochst. ex A. Rich.*) *Munro ex Benth.* = *Chloris simplex*
          Schumacher
339   Connarus thonningii *DC.* = *C. floribundus* Schum. et Thonn.
340   Indigofera secundiflora *Poir.* = *I. glutinosa* Schumacher
341   Panicum brevifolium *L.* = *P. plantagineum* Schumacher
341   Typha domingensis *Pers.* = *T. australis* Schum. et Thonn.
342   Aphania senegalensis (*Juss. ex Poir.*) *Radlk.* = *Ornithrophe thyrsoides* Baker
343   Diospyros ferrea (*Willd.*) *Bakh.* = *Ferreola guineensis* Schum. et Thonn.
343   Sporobolus virginicus (*L.*) *Kunth* = *Agrostis congener* Schumacher
344   Setaria pallide-fusca (*Schumacher*) *Stapf et Hubbard* = *Panicum pallide-fuscum*
          Schumacher
344   Pterocarpus santalinoides *L'Hèr. ex DC.* = *P. esculentus* Schum. et Thonn.
345   Heteropogon contortus (*L.*) *P. Beauv. ex Roem. et Schult.* = *Andropogon contortum* L.
346   Mariscus ligularis (*L.*) *Urb.* = *Cyperus ligularis* L.
347   Sclerocarpus africanus *Jacq. ex Murr.*
347   Kyllinga squamulata *Vahl*
348   Fimbristylis triflora (*L.*) *K. Schum.* = *Abildgaardia lanceolata* Schumacher
348   Borreria verticillata (*L.*) *G. F. W. Mey.* = *Spermacoce globosa* Schum. et Thonn.
349   Fimbristylis hispidula (*Vahl*) *Kunth* = *Scirpus hispidulus* Vahl
349   Aloe picta *Thunb.*
350   Sida alba *L.* = *S. scabra* Thonning
350?   Cyperus amabilis *Vahl.* = *C. microstachyos* Vahl
351   Blumea aurita (*L. f.*) *DC.* = *Erigeron stipulatum* Schum. et Thonn.
351   Eragrostis aspera (*Jacq.*) *Nees* = *Poa hippuris* Schumacher
352   Piliostigma thonningii (*Schumacher*) *Milne-Redh.* = *Bauhinia thonningii* Schumacher
352   Eragrostis ciliaris (*L.*) *R. Br.* = *Poa ciliaris* L.
353   Brachiaria lata (*Schumacher*) *C. E. Hubbard* = *Panicum latum* Schumacher
354   Paspalidium geminatum (*Forsk.*) *Stapf*
355   Pennisetum purpureum *Schumacher*
356   Aristida adscensionis *L.* = *A. submucronata* Schumacher
357   Aristida sieberana *Trin.* = *A. longiflora* Schumacher
358   Panicum maximum *Jacq.* = *P. sparsum* Schumacher
359   Vacant
360   Cenchrus biflorus *Roxb.* = *C. barbatus* Schumacher
361   Oplismenus hirtellus (*L.*) *P. Beauv.* = *Panicum incanum* Schumacher

362 Eragrostis turgida (*Schumacher*) *De Wild.* = *Poa turgida* Schumacher
363 Olyra latifolia *L.* = *O. brevifolia* Schumacher
364 Tricholaena monachne (*Trin.*) *Stapf et Hubbard* = *Aira bicolor* Schumacher
365 Dactyloctenium aegyptium (*L.*) *P. Beauv.* = *Chloris guineensis* Schumacher
366 Ctenium canescens *Benth.* = *Chloris pulchra* Schumacher
367 Digitaria horizontalis *Willd.* = *D. reflexa* Schumacher & *D. nuda* Schumacher
368 Vacant
369 Vacant
370 Cyperus maritims *Poir.* = *C. scirpoides* Vahl
371 Chloris pilosa *Schumacher*
372 Vacant
373 Brachiaria falcifera (*Trin.*) *Stapf* = *Panicum collare* Schumacher
374 Pennisetum polystachion (*L.*) *Schult.* = *Panicum cauda-ratti* Schumacher
375 Pennisetum subangustum (*Schumacher*) *Stapf et Hubbard* = *Panicum subangustum* Schumacher
376 Fimbristylis obtusifolia (*Lam.*) *Kunth* = *Scirpus obtusifolius* Lam.
377 Remirea maritima *Aubl.*
378 Bulbostylis barbata (*Rottb.*) *C. B. Cl.* = *Scirpus antarcticus* sensu Vahl
379 Kyllinga erecta *Schumacher*
380 Rhynchelytrum repens (*Willd.*) *C. E. Hubbard* = *Saccharum repens* Willd.
381 Sporobolus pyramidalis *P. Beauv.* = *Agrostis extensa* Schumacher
382 Eragrostis linearis (*Schumacher*) *Benth.* = *Poa linearis* Schumacher
383 Vacant
384 Mariscus alternifolius *Vahl*
385 Brachiaria distichophylla (*Trin.*) *Stapf* = *Panicum viviparum* Schumacher
386 Setaria barbata (*Lam.*) *Kunth* = *Panicum lineatum* Schumacher
387 Vacant
388 Cyperus bulbosus *Vahl* = *C. polyphyllus* Vahl
389 Cyperus margaritaceus *Vahl*
390 Brachiaria deflexa (*Schumacher*) *C. E. Hubbard* = *Panicum deflexum* Schumacher
391 Fimbristylis pilosa *Vahl*
392 Eleusine indica (*L.*) *Gaertn.* = *E. glabra* Schumacher
393 Bulbostylis pilosa (*Willd.*) *Cherm.* = *Schoenus pilosus* Willd.
394 Fimbristylis scabrida *Schumacher*
395 Brachiaria distichophylla (*Trin.*) *Stapf* = *Panicum serrulatum* Schumacher
396 Mariscus dubius (*Rottb.*) *C. E. C. Fischer* = *Cyperus coloratus* Vahl
397 Cyperus dilatatus *Schumacher*
398 Cyperus pustulatus *Vahl*
399 Cyperus sphacelatus *Rottb.*
400 Vacant
401 Setaria longiseta *P. Beauv.* = *Panicum longifolium* Schumacher
402 Paspalum orbiculare *Forsk.* = *Paspalus barbatus* Schumacher
403 Cyperus amabilis *Vahl*
404 Ascolepis dipsacoides (*Schumacher*) *J. Raynal* = *Kyllinga dipsacoides* Schumacher
405 Cyperus aristatus *Rottb.*
406 Hackelochloa granularis (*L.*) *Kuntze* = *Manisuris granularis* (L.) Sw.
407 Pycreus pumilus *L.* = *Cyperus patens* Vahl
408 Vacant
409 Vacant
410 Pycreus polystachyos (*Rottb.*) *P. Beauv.* = *Cyperus angustifolius* Schumacher

APPENDIX V

V(a)   INDEX TO MICROFICHE OF ISERT SPECIMENS TRACED IN THE WILLDENOW
HERBARIUM, BERLIN (B).

HERB. WILLD. NO.	CURRENT BOTANICAL NAME	NAME USED IN HERB. WILLD. INDEX (P. HIEPKO, 1972) IF DIFFERENT	MICROFICHE REF. I.D.C. 7440
47	Usteria guineensis *Willd.*		3:II.8
147	Dialium guineense *Willd.*	*Codarium nitidum* Sol. ex Vahl (2 sheets)	9:III.3,4
768	Boerhavia diffusa *L.*	*B. adscendens* Willd.	46:II.6
860	Hippocratea africana (*Willd.*) Loes. ex Engl.	*Tonsella africana* Willd.	52:I.5
1095	Bulbostylis pilosa (*Willd.*) Cherm	*Schoenus pilosus* Willd.	63:I.2
1291	Cyperus margaritaceus *Vahl*		75:I.4
1499	Rhynchelytrum repens (*Willd.*) C. E. Hubbard	*Saccharum repens* Willd.	87:III.7
2596	Kohautia virgata (*Willd.*) Bremek.	*Hedyotis virgata* Willd.	173:I.5
3046	Fagara zanthoxyloides *Lam.*		204:II.1
3253	Heliotropium strigosum *Willd.*		219:II.7
3630	Hewittia sublobata (*L.f.*) *Kuntze*	*Convolvulus involucratus* Willd.	250:III.8
3638	Merremia tridentata (*L.*) *Hallier* f. subsp. angustifolia (*Jacq.*) Oostr.	*Ipomoea angustifolia* Jacq. (2 sheets)	251:II.3,4
3746	Ipomoea coptica (*L.*) *Roth ex Roem. et Schult.*	*I. dissecta* Willd.	258:II.3
3908	Mitragyna inermis (*Willd.*) Kuntze	*Nauclea inermis* Willd.	274:II.5
4997	Polycarpaea stellata (*Willd.*) DC.	*Mollia stellata* (Willd.) Willd.	342:III.3
5271	Pergularia daemia (*Forsk.*) Chiov.	*Asclepias convolvulacea* Willd.	360:I.3
5338	Chenopodium murale *L.*	*Chenopodium guineense* Jacq.	363:II.9
5357	Amaranthus viridis *L.*	*Chenopodium caudatum* Jacq.	365:I.5,6
(6711	Sansevieria liberica *Gér. et Labr.*	*S. guineensis* ? Isert's	466:II.1)
7107	Limnophyton obtusifolium (*L.*) Miq.	*Alisma sagittifolium* Willd.	492*:I.4
7126	Pancovia bijuga *Willd.* (2 sheets)		491*:II.5,6
7249	Clappertonia ficifolia (*Willd.*) Decne.	*Honkenya ficifolia* Willd.	500:II.2
7292	Clausena anisata (*Willd.*) Hook. f. ex Benth.	*Amyris anisata* Willd.	502:II.5
7732	Cardiospermum grandiflorum *Sw.*	*C. hirsutum* Willd.	533:II.1
8131	Ludwigia octovalvis (*Jacq.*) Raven subsp. brevisepala (Brenan) Raven	*Jussieua linearis* Willd.	555:III.2

* These microfiches have been incorrectly numbered: 491 should be 492 and vice versa.

HERB. WILLD. NO.	CURRENT BOTANICAL NAME	NAME USED IN HERB. WILLD. INDEX (P. HIEPKO, 1972) IF DIFFERENT	MICROFICHE REF. I.D.C. 7440
8861	Triaspis odorata (*Willd.*) A. *Juss.*	*Hiraea odorata* Willd.	614:II.7
9582	Syzygium guineense (*Willd.*) DC. var. guineense	*Calyptranthes guineensis* Willd.	660:II.6
10039	Capparis erythrocarpos *Isert*		698:II.6
10353	Tetracera alnifolia *Willd.* (2 sheets)		722:III.5,6
10972	Phaulopsis ciliata (*Willd.*) *Hepper*	*Origanum ciliatum* Willd.	777:III.2
11182	Striga aspera (*Willd.*) Benth.	*Euphrasia aspera* Willd.	796:I.4
11658	Hygrophila auriculata (*Schum. et Thonn.*) *Heine*	*Barleria longifolia* L.	829:III.7
11682	Clerodendrum capitatum (*Willd.*) *Schum. et Thonn.* var capitatum	*Volkameria capitata* Willd.	831:II.7
12977	Polygala arenaria *Willd.*		930:III.2
12980	Polygala guineensis *Willd.*		930:III.6
13242	Crotalaria glauca *Willd.*		945:II.6
13411	Pseudovigna argentea (*Willd.*) *Verdc.*	*Dolichos argenteus* Willd.	957:II.4
(13432	Galactia rubra (*Jacq.*) *Urb.*	*G. sericea* Willd.	958:III.1)
13437	Ophrestia hedysaroides (*Willd.*) *Verdc.*	*Glycine hedysaroides* Willd.	958:III.9
13446	Atylosia scarabaeoides (*L.*) *Benth.*	*Glycine mollis* Willd.	959:I.9
13645	Ormocarpum sennoides (*Willd.*) D.C. subsp. hispidum (*Willd.*) Brenan et *J. Léonard*	*Cytisus hispidus* Willd.	974:II.6
13757	Alysicarpus rugosus (*Willd.*) DC.	*Hedysarum rugosum* Willd.	983:III.8
13834	Uraria picta (*Jacq.*) DC.	*Hedysarum pictum* Jacq. (2 sheets)	987:III.9; 988.I.1
13903	Indigofera dendroides *Jacq.*		994:I.4
13911	Indigofera pulchra *Willd.* (3 sheets)		994:II.9, III.1,2
13941	Tephrosia linearis (*Willd.*) *Pers.*	*Galega linearis* Willd.	996:II.3
14357	Citropsis articulata (*Willd. ex Spreng.*) *Swingle et Kellermann*	*Citrus articulata* Willd. ex Spreng.	1027:II.7
14538	Launaea taraxacifolia (*Willd.*) *Amin ex C. Jeffrey*	*Sonchus taraxacifolia* Willd.	1039:III.5
15275	Enydra fluctuans *Lour.*	*Caesulia radicans* Willd.	1097:I.7
15668	Blumea aurita (*L.f.*) DC.	*Conyza guineensis* Willd.	1124:III.1
17317	Scleria verrucosa *Willd.*		1251:I.5
17807	Acalypha ciliata *Forsk.*		1289:II.3
17910	Croton lobatus *L.*	*C. trilobatum* Forsk.	1295:I.7
18018	Kedrostis foetidissima (*Jacq.*) *Cogn.*	*Trichosanthes foetidissima* Jacq. (2 sheets)	1301:III.2,3
18030	Lagenaria siceraria (*Molina*) *Standley*	*Cucurbita idolatrica* Willd.	1302:II.7
18491	Flacourtia flavescens *Willd.*		1344:II.7
18639	Sorghum arundinaceum (*Desv.*) *Stapf*	*Andropogon arundinaceus* Willd.	1354:III.6

HERB. WILLD. NO.	CURRENT BOTANICAL NAME	NAME USED IN HERB. WILLD. INDEX (P. HIEPKO, 1972) IF DIFFERENT	MICROFICHE REF. I.D.C. 7440
19083	Mimosa pigra *L.*	*Mimosa canescens* Willd.	1384:I.5
19312	Ficus exasperata *Vahl*	*Ficus scabra* Willd.	1399:III.5
19587	Blechnum seminudum *Willd.*	*Grammitis seminuda* (Willd.) Willd.	1416:II.1

V(b) THONNING'S SPECIMEN TRACED IN THE BRITISH MUSEUM (NATURAL HISTORY) (BM).

Thonningia sanguinea *Vahl*
[Afzelia parviflora (*Vahl*) *Hepper*—Smeathman specimen (see p. 33)].

V(c) INDEX TO MICROFICHE OF ISERT AND THONNING SPECIMENS IN THE BOTANICAL MUSEUM, COPENHAGEN (C).

REFERENCES TO INDIVIDUAL MICROFICHE FRAMES AND THEIR CURRENT BOTANICAL IDENTIFICATION.

IDC microfiche
No. 2203

1:	I. 1–2	Fimbristylis triflora (*L.*) *K. Schum.*
	3–7	Mallotus oppositifolius (*Geisel.*) *Müll. Arg.*
	II. 1–2	Mallotus oppositifolius (*Geisel.*) *Müll. Arg.*
	3–4	Amyris sylvatica *Jacq.*
	5–7	Acalypha ciliata *Forsk.*
	III. 1–7	Acalypha ciliata *Forsk.*
2:	I. 1–2	Acalypha ciliata *Forsk.*
	3–6	Chrysophyllum albidum *G. Don*
	7	Pupalia lappacea (*L.*) *Juss.*
	II. 1–5	Pupalia lappacea (*L.*) *Juss.*
	6	Acrostichum aureum *L.*
	7	Tetrapleura tetraptera (*Schum. et Thonn.*) *Taub.*
	III. 1–3	Tetrapleura tetraptera (*Schum. et Thonn.*) *Taub.*
	4–7	Doryopteris kirkii (*Hook.*) *Alston*
3:	I. 1–5	Doryopteris kirkii (*Hook.*) *Alston*
	6–7	Arthropteris palisotii *Desv.*
	II. 1	Arthropteris palisotii *Desv.*
	2	Aeschynomene sensitiva *Sw.*
	3–6	Aeschynomene indica *L.*
	7	Agelaea obliqua (*P. Beauv.*) *Baill.*
	III. 1–7	Sporobolus virginicus (*L.*) *Kunth*
4:	I. 1–2	Sporobolus virginicus (*L.*) *Kunth*
	3–7	Sporobolus pyramidalis *P. Beauv.*
	II. 1–5	Sporobolus pyramidalis *P. Beauv.*
	6	Tricholaena monachne (*Trin.*) *Stapf et C. E. Hubbard*
	7	Amaranthus viridis *L.*
	III. 1–3	Amaranthus viridis *L.*
	4–5	Amaranthus lividus *L.*
	6–7	Cyathula prostrata (*L.*) *Blume*

IDC microfiche
No. 2203

5:	I. 1–2	Amaranthus viridis *L.*
	3	Pancratium hirtum *A. Chev.*
	4	Crinum jagus (*Thomps.*) *Dandy*
	5	Costus afer *Ker-Gawl.*
	6–7	Aframomum sceptrum (*Oliv. et Hanb.*) *K. Schum.*
	II. 1–5	Aframomum sceptrum (*Oliv. et Hanb.*) *K. Schum.*
	6	Aframomum melegueta *K. Schum.*
	7	Sorindeia warneckei *Engl.*
	III. 1–2	Rutaceae: non West African
	3–7	Clausenia anisata (*Willd.*) *Hook. f. ex Benth.*
6:	I. 1–3	Clausenia anisata (*Willd.*) *Hook. f. ex Benth.*
	4	Allophylus spicatus (*Poir.*) *Radlk.*
	5–7	Ancistrophyllum secundiflorum (*P. Beauv.*) *Wendl.*
	II. 1	Anthephora cristata (*Doell*) *Hack. ex De Wild. et Dur.*
	2	Aponogeton subconjugatus *Schum. et Thonn.*
	3–5	Arachis hypogaea *L.*
	6	Argemone mexicana *L.*
	7	Aristida adscensionis *L.*
	III. 1–3	Aristida adscensionis *L.*
	4–7	Aristida sieberana *Trin.*
7:	I. 1	Arthropteris palisotii *Desv.*
	2–7	Pergularia daemia (*Forsk.*) *Chiov.*
	II. 1–7	Pergularia daemia (*Forsk.*) *Chiov.*
	III. 1–7	Pergularia daemia (*Forsk.*) *Chiov.*
8:	I. 1	Sarcostemma viminale (*L.*) *R. Br.*
	2–7	Calotropis procera (*Ait.*) *Ait. f.*
	II. 1–2	Asparagus warneckei (*Engl.*) *Hutch.*
	3–5	Ctenitis cirrhosa (*Schumacher*) *Ching*
	6–7	Nephrolepis biserrata (*Sw.*) *Schott*
	III. 1–2	Nephrolepis biserrata (*Sw.*) *Schott*
	3	Cyclosorus striatus (*Schumacher*) *Ching*
	4	Arthropteris orientalis (*Gmel.*) *Posth.*
	5–7	Cyclosorus dentatus (*Forsk.*) *Ching*
9:	I. 1–7	Cyclosorus dentatus (*Forsk.*) *Ching*
	II. 1–2	Cyclosorus dentatus (*Forsk.*) *Ching*
	3–7	Nephrolepis biserrata (*Sw.*) *Schott*
	III. 1–7	Nephrolepis biserrata (*Sw.*) *Schott*
10:	I. 1–3	Nephrolepis biserrata (*Sw.*) *Schott*
	4	Asplenium africanum *Desv.*
	5–7	Avicennia germinans (*L*). *L.*
	II. 1–2	Azolla africana *Desv.*
	3–5	Hygrophila auriculata (*Schum. et Thonn.*) *Heine*
	6–7	Newboldia laevis (*P. Beauv.*) *Seemann ex Bureau*
	III. 1	Newboldia laevis (*P. Beauv.*) *Seemann ex Bureau*
	2	Spathodea campanulata *P. Beauv.*
	3–6	Blepharis maderaspatensis (*L.*) *Heyne ex Roth* subsp. rubiifolia (*Schum. et Thonn.*) *Napper*
11:	I. 1	Blepharis maderaspatensis (*L.*) *Heyne ex Roth* subsp. rubiifolia (*Schum. et Thonn.*) *Napper*
	2–5	Boerhavia diffusa *L.*
	6–7	Ceiba pentandra (*L.*) *Gaertn.*
	II. 1–4	Bridelia ferruginea *Benth.*
	5–7	Striga aspera (*Willd.*) *Benth.*

IDC microfiche
No. 2203

11: III.	1–3	Striga linearifolia (*Schum. et Thonn.*) *Hepper*
	4–7	Synsepalum dulcificum (*Schum. et Thonn.*) *Daniell*
12: I.	1–4	Synsepalum dulcificum (*Schum. et Thonn.*) *Daniell*
	5–7	Melanthera scandens (*Schum. et Thonn.*) *Roberty*
II.	1	Melanthera scandens (*Schum. et Thonn.*) *Roberty*
	2–5	Byrsocarpus puniceus *Schum. et Thonn.*
	6–7	Byrsocarpus coccineus *Schum. et Thonn.*
III.	1–7	Byrsocarpus coccineus *Schum. et Thonn.*
13: I.	1	Byrsocarpus coccineus *Schum. et Thonn.*
	2–6	Hyptis pectinata (*L.*) *Poit.*
	7	Crassocephalum rubens (*Juss. ex Jacq.*) *S. Moore*
II.	1–5	Crassocephalum rubens (*Juss. ex Jacq.*) *S. Moore*
	6–7	Caesalpinia pulcherrima (*L.*) *Sw.*
III.	1–2	Caesalpinia pulcherrima (*L.*) *Sw.*
	3	Syzygium guineense (*Willd.*) *DC.* var. guineense
	4–5	Canna indica *L.*
14: I.	1–3	Capparis erythrocarpos *Isert*
	4–6	Ritchiea reflexa (*Thonning*) *Gilg et Benedict*
	7	Capparis brassii *DC.*
II.	1–4	Capparis brassii *DC.*
	5	Capsicum frutescens *L.*
	6–7	Cardiospermum halicacabum *L.*
III.	1–2	Cardiospermum halicacabum *L.*
	3–4	Cardiospermum grandiflorum *Sw.*
	5–7	Carissa edulis *Vahl*
15: I.	1–6	Carissa edulis *Vahl*
	7	Cassia podocarpa *Guill. et Perr.*
II.	1–2	Cassia occidentalis *L.*
	3–7	Cassia mimosoides *L.*
III.	1–4	Cassia mimosoides *L.*
	5–6	Cassia absus *L.*
16: I.	1–6	Cassia absus *L.*
	7	Cassytha filiformis *L.*
II.	1–6	Cassytha filiformis *L.*
	7	Maytenus undata (*Thunb.*) *Blakelock*
III.	1–4	Maytenus undata (*Thunb.*) *Blakelock*
	5–6	Celosia trigyna *L.*
17: I.	1	Celosia isertii *C. C. Townsend*
	2	Celosia trigyna *L.*
	3–7	Celosia argentea *L.*
II.	1–3	Celosia argentea *L.*
	4–7	Trema orientalis (*L.*) *Blume*
III.	1–7	Trema orientalis (*L.*) *Blume*
18: I.	1–2	Trema orientalis (*L.*) *Blume*
	3–4	Cenchrus biflorus *Roxb.*
	5–7	Nauclea latifolia *Sm.*
II.	1	Raphionacme brownii *Scott Elliot*
	2	Morinda lucida *Benth.*
	3	Chenopodium murale *L.*
	4–7	Dactyloctenium aegyptium (*L.*) *P. Beauv.*
III.	1–3	Dactyloctenium aegyptium (*L.*) *P. Beauv.*
	4–7	Chloris pilosa *Schumacher*
19: I.	1–7	Enteropogon macrostachyus (*Hochst. ex A. Rich.*) *Munro ex Benth.*

IDC microfiche
No. 2203

19:	II. 1	Enteropogon macrostachyus (*Hochst. ex A. Rich.*) *Munro ex Benth.*
	2–7	Ctenium canescens *Benth.*
	III. 1	Ctenium canescens *Benth.*
	2–7	Chrysobalanus orbicularis *Schumacher*
20:	I. 1–5	Chrysobalanus orbicularis *Schumacher*
	6–7	Vernonia colorata (*Willd.*) *Drake*
	II. 1–2	Vernonia cinerea (*L.*) *Less.*
	3	Cissus quadrangularis *L.*
	4–6	Cyphostemma cymosa (*Schum. et Thonn.*) *Descoings*
	7	Cissus quadrangularis *L.*
	III. 1	Cissus quadrangularis *L.*
	2–7	Afraegle paniculata (*Schum. et Thonn.*) *Engl.*
21:	I. 1–5	Cleome rutidosperma *DC.*
	6	Gynandropsis gynandra (*L.*) *Briq.*
	7	Clerodendrum capitatum (*Willd.*) *Schum. et Thonn.* var. capitatum
	II. 1–2	Clerodendrum capitatum (*Willd.*) *Schum. et Thonn.* var. capitatum
	3–6	Bridelia ferruginea *Benth.*
	7	Cnestis ferruginea *DC.*
	III. 1–6	Cnestis ferruginea *DC.*
	7	Dialium guineense *Willd.*
22:	I. 1–2	Dialium guineense *Willd.*
	3–6	Coldenia procumbens *L.*
	7	Combretum platypterum (*Welw.*) *Hutch. et Dalz.*
	II. 1–7	Murdannia simplex (*Vahl*) *Brenan*
	III. 1	Murdannia simplex (*Vahl*) *Brenan*
	2–7	Commelina erecta *L.*
23:	I. 1–4	Aneilema umbrosum (*Vahl*) *Kunth*
	5–7	Connarus thonningii (*DC.*) *Schellenb.*
	II. 1–5	Connarus thonningii (*DC.*) *Schellenb.*
	6–7	Connarus monocarpus *L.*
	III. 1–2	Connarus monocarpus *L.*
	3	Connarus africanus *Lam.*
	4–6	Conocarpus erectus *L.*
24:	I. 1–3	Ipomoea involucrata *P. Beauv.*
	4–5	Ipomoea pes-caprae (*L.*) *Sweet* subsp. brasiliensis (*L.*) *Oostr.*
	6–7	Ipomoea ochracea (*Lindl.*) *G. Don*
	II. 1	Ipomoea ochracea (*Lindl.*) *G. Don*
	2–5	Ipomoea stolonifera (*Cyrill.*) *J. E. Gmel.*
	6–7	Hewittia sublobata (*L.f.*) *Kuntze*
	III. 1–3	Hewittia sublobata (*L.f.*) *Kuntze*
	4–5	Ipomoea mauritiana *Jacq.*
	6–7	Merremia aegyptiaca (*L.*) *Urban*
25:	I. 1–3	Convolvulus cairica *L.*
	4–7	Ipomoea cairica (*L.*) *Sweet*
	II. 1–3	Jacquemontia ovalifolia (*Vahl*) *Hall.f.*
	4–7	Ipomoea sepiaria *Roxb.*
	III. 1–3	Jacquemontia tamnifolia (*L.*) *Griseb.*
	4–7	Blumea aurita (*L.f.*) *DC.*
26:	I. 1–7	Corchorus tridens *L.*
	II. 1–2	Corchorus tridens *L.*
	3–7	Corchorus trilocularis *L.*
	III. 1–4	Corchorus trilocularis *L.*
	5–7	Corchorus aestuans *L.*

IDC microfiche
No. 2203

27:   I. 1   Corchorus aestuans *L.*
      2–3   Cordia guineensis *Thonning*
      4–5   Aspilia helianthoides (*Schum. et Thonn.*) *Oliv. et Hiern*
      6–7   Costus afer *Ker-Gawl.*
     II. 1–7   Crotalaria goreensis *Guill. et Perr.*
    III. 1   Crotalaria goreensis *Guill. et Perr.*
      2–6   Crotalaria glauca *Willd.*
28:   I. 1–4   Crotalaria glauca *Willd.*
      5–7   Crotalaria pallida *Ait.* var. obovata (*G. Don*) *Polhill*
     II. 1–3   Crotalaria pallida *Ait.* var. obovata (*G. Don*) *Polhill*
      4–6   Crotalaria pallida *Ait.* var. pallida
      7   Mallotus oppositifolius (*Geisel.*) *Müll. Arg.*
    III. 1–2   Mallotus oppositifolius (*Geisel.*) *Müll. Arg.*
      3–4   Croton lobatus *L.*
      5–7   Blighia sapida *König*
29:   I. 1–4   Blighia sapida *König*
      5–7   Leptadenia hastata (*Pers.*) *Decne.*
     II. 1–7   Leptadenia hastata (*Pers.*) *Decne.*
    III. 1–7   Leptadenia hastata (*Pers.*) *Decne.*
30:   I. 1–2   Leptadenia hastata (*Pers.*) *Decne.*
      3–7   Gymnema sylvestre (*Retz.*) *Schultes*
     II. 1–2   Gymnema sylvestre (*Retz.*) *Schultes*
      3–6   Cynometra vogelii *Hook. f.*
      7   Cyperus amabilis *Vahl*
    III. 1–7   Cyperus amabilis *Vahl*
31:   I. 1–4   Pycreus polystachyos (*Rottb.*) *P. Beauv.*
      5–7   Cyperus aristatus *Rottb.*
     II. 1   Cyperus aristatus *Rottb.*
      2–3   Cyperus articulatus *L.*
      4–7   Mariscus dubius (*Rottb.*) *C. E. C. Fischer*
    III. 1–4   Cyperus difformis *L.*
      5–7   Cyperus dilatatus *Schumacher*
32:   I. 1–2   Cyperus dilatatus *Schumacher*
      3–4   Cyperus distans *L. f.*
      5–7   Mariscus ligularis (*L.*) *Urb.*
     II. 1–2   Mariscus ligularis (*L.*) *Urb.*
      3–7   Cyperus margaritaceus *Vahl*
    III. 1–7   Cyperus margaritaceus *Vahl*
33:   I. 1–6   Cyperus amabilis *Vahl*
      7   Pycreus pumilus *L.*
     II. 1–3   Pycreus pumilus *L.*
      4–7   Cyperus nudicaulis *Poir.*
    III. 1–7   Cyperus bulbosus *Vahl*
34:   I. 1–4   Cyperus pustulatus *Vahl*
      5–7   Cyperus imbricatus *Retz.*
     II. 1   Cyperus imbricatus *Retz.*
      2–5   Cyperus maritimus *Poir.*
      6–7   Cyperus sphacelatus *Rottb.*
    III. 1–4   Cyperus sphacelatus *Rottb.*
      5   Cyperus fenzelianus *Steud.*
      6   Cytosperma senegalense (*Schott*) *Engl.*
      7   Cajanus cajan (*L.*) *Millsp.*
35:   I. 1–2   Cajanus cajan (*L.*) *Millsp.*

IDC microfiche
No. 2203

35:	I. 3–4	Datura metel *L.*
	5–7	Deinbollia pinnata (*Poir.*) *Schum. et Thonn.*
	II. 1–7	Deinbollia pinnata (*Poir.*) *Schum. et Thonn.*
	III. 1–3	Deinbollia pinnata (*Poir.*) *Schum. et Thonn.*
	4–7	Dialium guineense *Willd.*
36:	I. 1	Digitaria horizontalis *Willd.*
	2	Digitaria leptorhachis (*Pilger*) *Stapf*
	3	Digitaria horizontalis *Willd.*
	4–5	Diodia sarmentosa *Sw.*
	6–7	Borreria scabra (*Schum. et Thonn.*) *K. Schum.*
	II. 1	Borreria scabra (*Schum. et Thonn.*) *K. Schum.*
	2–7	Diplazium proliferum (*Lam.*) *Kaulf.*
	III. 1–7	Diplazium proliferum (*Lam.*) *Kaulf.*
37:	I. 1–4	Dodonaea viscosa *Jacq.*
	5	Lablab purpureus (*L.*) *Sweet*
	6–7	Canavalia rosea (*Sw.*) *DC.*
	II. 1–3	Canavalia rosea (*Sw.*) *DC.*
	4–5	Phaseolus adenanthus *G. F. W. Mey.*
	6	Dioclea reflexa *Hook. f.*
	7	Canavalia virosa (*Roxb.*) *Wight et Arn.*
	III. 1	Dioclea reflexa *Hook. f.*
	2–3	Vigna luteola (*Jacq.*) *Benth.*
	4–5	Canavalia virosa (*Roxb.*) *Wight et Arn.*
	6	Echinochloa pyramidalis (*Lam.*) *Hitchc. et Chase*
	7	Motandra guineensis (*Thonning*) *A. DC.*
38:	I. 1–7	Motandra guineensis (*Thonning*) *A. DC.*
	II. 1	Motandra guineensis (*Thonning*) *A. DC.*
	2–5	Spilanthes filicaulis (*Schum. et Thonn.*) *C. D. Adams*
	6–7	Eclipta alba (*L.*) *Hassk.*
	III. 1–4	Eclipta alba (*L.*) *Hassk.*
	5–7	Ehretia corymbosa *Bojer ex A. DC.*
39:	I. 1–7	Ehretia corymbosa *Bojer ex A. DC.*
	II. 1–3	Ehretia corymbosa *Bojer ex A. DC.*
	4	Eleocharis acutangula (*Roxb.*) *Schult.*
	5–7	Sesbania pachycarpa *DC. emend. Guill. et Perr.*
	III. 1–5	Sesbania pachycarpa *DC. emend. Guill. et Perr.*
	6–7	Sesbania sericea (*Willd.*) *Link*
40:	I. 1–3	Sesbania sericea (*Willd.*) *Link*
	4–5	Conyza aegyptiaca (*L.*) *Ait.*
	6–7	Erythrina senegalensis *DC.*
	II. 1–4	Erythroxylum emarginatum *Thonning*
	5–7	Eugenia coronata *Schum. et Thonn.*
	III. 1	Eugenia coronata *Schum. et Thonn.*
	2	Eugenia jambos *L.*
	3	Syzygium guineense (*Willd.*) *DC.* var. littorale *Keay*
	4–7	Euphorbia prostrata *Ait.*
41:	I. 1–4	Euphorbia lateriflora *Schum. et Thonn.*
	5–7	Euphorbia purpurascens *Schum. et Thonn.*
	II. 1–7	Euphorbia purpurascens *Schum. et Thonn.*
	III. 1–4	Euphorbia glaucophylla *Poir.*
	5–7	Evolvulus alsinoides (*L.*) *L.*
42:	I. 1–3	Evolvulus alsinoides (*L.*) *L.*
	4–5	Diospyros tricolor (*Schum. et Thonn.*) *Hiern*

IDC microfiche
No. 2203

42 :	I. 6–7	Diospyros ferrea (*Willd.*) *Bakh.*
	II. 1–3	Diospyros ferrea (*Willd.*) *Bakh.*
	4	Ficus exasperata *Vahl*
	5	Ficus polita *Vahl*
	6–7	Fimbristylis pilosa *Vahl*
	III. 1–4	Fimbristylis pilosa *Vahl*
	5–6	Fimbristylis scabrida *Schumacher*
	7	Flacourtia flavescens *Willd.*
43 :	I. 1–5	Flagellaria guineensis *Schumacher*
	6–7	Fuirena umbellata *Rottb.*
	II. 1–2	Fuirena umbellata *Rottb.*
	3–7	Tephrosia linearis (*Willd.*) *Pers.*
	III. 1	Tephrosia linearis (*Willd.*) *Pers.*
	2	Tephrosia bracteolata *Guill. et Perr.*
	3–7	Gardenia ternifolia *Schum. et Thonn.*
44 :	I. 1–2	Gardenia ternifolia *Schum. et Thonn.*
	3–4	Rutidea parviflora *DC.*
	5–7	Micrargeria filiformis (*Schum. et Thonn.*) *Hutch. et Dalz.*
	II. 1	Micrargeria filiformis (*Schum. et Thonn.*) *Hutch. et Dalz.*
	2–7	Curculigo pilosa (*Schum. et Thonn.*) *Engl.*
	III. 1–4	Curculigo pilosa (*Schum. et Thonn.*) *Engl.*
	5–7	Gisekia pharnaceoides *L.*
45 :	I. 1–2	Gisekia pharnaceoides *L.*
	3	Gloriosa superba *L.*
	4–7	Macrotyloma biflorum (*Schum. et Thonn.*) *Hepper*
	II. 1–4	Macrotyloma biflorum (*Schum. et Thonn.*) *Hepper*
	5–7	Ophrestia hedysaroides (*Willd.*) *Verdc.*
	III. 1–6	Ophrestia hedysaroides (*Willd.*) *Verdc.*
	7	Rhynchosia pycnostachya (*DC.*) *Meikle*
46 :	I. 1–6	Rhynchosia pycnostachya (*DC.*) *Meikle*
	7	Rhynchosia minima (*L.*) *DC.*
	II. 1–4	Rhynchosia minima (*L.*) *DC.*
	5–7	Eriosema glomeratum (*Guill. et Perr.*) *Hook. f.*
	III. 1–2	Eriosema glomeratum (*Guill. et Perr.*) *Hook. f.*
	3–4	Galactia tenuiflora (*Willd.*) *Wight et Arn.*
	5–7	Rhynchosia sublobata (*Schum. et Thonn.*) *Meikle*
47 :	I. 1–6	Rhynchosia sublobata (*Schum. et Thonn.*) *Meikle*
	7	Eleusine indica (*L.*) *Gaertn.*
	II. 1–2	Eleusine indica (*L.*) *Gaertn.*
	3–6	Ouratea flava (*Schum. et Thonn.*) *Hutch. et Dalz. ex Stapf*
	7	Ochna multiflora *DC.*
	III. 1	Ochna membranacea *Oliv.*
	2	Spondias mombin *L.*
	3	Ochna multiflora *DC.*
	4–7	Pandiaka angustifolia (*Vahl*) *Hepper*
48 :	I. 1–4	Pandiaka angustifolia (*Vahl*) *Hepper*
	5–6	Philoxerus vermicularis (*L.*) *P. Beauv.*
	7	Grewia carpinifolia *Juss.*
	II. 1–7	Grewia carpinifolia *Juss.*
	III. 1–2	Grewia carpinifolia *Juss.*
	3–6	Hoslundia opposita *Vahl et Thonn.*
49 :	I. 1–2	Hoslundia opposita *Vahl et Thonn.*
	3	Haemanthus multiflorus *Martyn*

IDC microfiche
No. 2203

49: I. 4–7   Urera mannii (*Wedd.*) *Benth. et Hook. f. ex Rendle*
     II. 1–2   Urera mannii (*Wedd.*) *Benth. et Hook. f. ex Rendle*
         3–4   Laportea ovalifolia (*Schum. et Thonn.*) *Chew*
         5–7   Oldenlandia lancifolia (*Schumacher*) *DC.*
     III. 1–3 Oldenlandia corymbosa *L.*
         4–5   Oldenlandia lancifolia (*Schumacher*) *DC.*
         6–7   Pentodon pentandrus (*Schum. et Thonn.*) *Vatke*
50: I. 1–7   Pentodon pentandrus (*Schum. et Thonn.*) *Vatke*
     II. 1   Pentodon pentandrus (*Schum. et Thonn.*) *Vatke*
         2–7   Kohautia virgata (*Willd.*) *Bremek.*
     III. 1–4 Desmodium velutinum (*Willd.*) *DC.*
         5–7   Desmodium ramosissimum *G. Don*
51: I. 1–2   Desmodium ramosissimum *G. Don*
         3–7   Desmodium triflorum (*L.*) *DC.*
     II. 1–3   Desmodium gangeticum (*L.*) *DC.*
         4–7   Alysicarpus ovalifolius (*Schum. et Thonn.*) *J. Léonard*
     III. 1–3 Alysicarpus ovalifolius (*Schum. et Thonn.*) *J. Léonard*
         4–6   Uraria picta (*Jacq.*) *DC.*
52: I. 1–4   Desmodium velutinum (*Willd.*) *DC.*
         5   Heliotropium indicum *L.*
         6–7   Heliotropium strigosum *Willd.*
     II. 1–7   Heliotropium strigosum *Willd.*
     III. 1   Iphigenia ledermannii *Engl. et K. Krause*
         2   Hibiscus sp.
         3   Hibiscus micranthus *L. f.*
         4–5   Abelmoschus moschatus *Medik.*
         6–7   Hibiscus cannabinus *L.*
53: I. 1–4   Hibiscus cannabinus *L.*
         5–6   Hibiscus esculentus *L.*
         7   Hibiscus cannabinus *L.*
     II. 1–2   Hibiscus cannabinus *L.*
         3–7   Hibiscus vitifolius *L.*
     III. 1–5 Hibiscus surattensis *L.*
         6–7   Hibiscus owariensis *P. Beauv.*
54: I. 1–6   Hibiscus owariensis *P. Beauv.*
         7   Hibiscus micranthus *L.*
     II. 1   Hibiscus micranthus *L.*
         2–5   Triaspis odorata (*Willd.*) *A. Juss.*
         6–7   Clappertonia ficifolia (*Willd.*) *Decne.*
     III. 1   Clappertonia ficifolia (*Willd.*) *Decne.*
         2–4   Hydrolea glabra *Schum. et Thonn.*
         5–7   Lipocarpha sphacelata (*Vahl*) *Nees*
55: I. 1–2   Lipocarpha sphacelata (*Vahl*) *Nees*
         3   Hyperthelia dissoluta (*Nees ex Steud.*) *W. D. Clayton*
         4–7   Hyptis lanceolata *Poir.*
     II. 1–4   Hyptis lanceolata *Poir.*
         5–6   Hyptis pectinata (*L.*) *Poit.*
         7   Aerva lanata (*L.*) *Juss. ex Schult.*
     III. 1–7 Aerva lanata (*L.*) *Juss. ex Schult.*
56: I. 1–4   Alternanthera pungens *Kunth*
         5–7   Alternanthera sessilis (*L.*) *DC.*
     II. 1–3   Alternanthera sessilis (*L.*) *DC.*
         4   Imperata cylindrica (*L.*) *P. Beauv.* var. africana (*Anders.*) *C .E. Hubbard*

IDC microfiche
No. 2203

56:	II. 5–7	Indigofera secundiflora *Poir.*
	III. 1–2	Indigofera secundiflora *Poir.*
	3–7	Indigofera pilosa *Poir.*
57:	I. 1–7	Indigofera paniculata *Vahl ex Pers.*
	II. 1–4	Indigofera pulchra *Willd.*
	5–7	Indigofera subulata *Vahl ex Poir.*
	III. 1–4	Indigofera subulata *Vahl ex Poir.*
	5–6	Indigofera aspera *Perr. ex DC.*
	7	Indigofera tetrasperma *Vahl ex Pers.*
58:	I. 1–7	Indigofera tetrasperma *Vahl ex Pers.*
	II. 1–2	Indigofera tetrasperma *Vahl ex Pers.*
	3–7	Indigofera nigricans *Vahl ex Pers.*
	III. 1–2	Indigofera nigricans *Vahl ex Pers.*
	3–5	Indigofera macrophylla *Schumacher*
	6–7	Indigofera hirsuta *L.*
59:	I. 1–2	Indigofera hirsuta *L.*
	3–7	Indigofera dendroides *Jacq.*
	II. 1–7	Indigofera dendroides *Jacq.*
	III. 1–7	Indigofera spicata *Forsk.*
60:	I. 1–6	Indigofera spicata *Forsk.*
	7	Indigofera tinctoria *L.*
	II. 1	Indigofera tinctoria *L.*
	2	Ipomoea ochracea (*Lindl.*) *G. Don*
	3	Aniseia martinicensis (*Jacq.*) *Choisy*
	4–7	Merremia tridentata (*L.*) *Hall. f.* subsp. angustifolia (*Jacq.*) *Oostr.*
	III. 1–5	Merremia tridentata (*L.*) *Hall. f.* subsp. angustifolia (*Jacq.*) *Oostr.*
	6–7	Ipomoea coptica (*L.*) *Roth ex Roem. et Schult.*
61:	I. 1–3	Ipomoea coptica (*L.*) *Roth ex Roem. et Schult.*
	4–7	Ipomoea heterotricha *F. Didr.*
	II. 1	Pavetta corymbosa (*DC.*) *F. N. Williams*
	2–6	Jasminum dichotomum *Vahl*
	7	Jatropha curcas *L.*
	III. 1–3	Jatropha curcas *L.*
	4–7	Ludwigia octovalvis (*Jacq.*) *Raven* subsp. brevisepala (*Brenan*) *Raven*
62:	I. 1–2	Ludwigia octovalvis (*Jacq.*) *Raven* subsp. brevisepala (*Brenan*) *Raven*
	3–5	Justicia flava (*Forsk.*) *Vahl*
	6–7	Ruspolia hypocrateriformis (*Vahl*) *Milne-Redh.*
	II. 1–7	Ruspolia hypocrateriformis (*Vahl*) *Milne-Redh.*
	III. 1–2	Barleria opaca (*Vahl*) *Nees*
	3–6	Ascolepis dipsacoides (*Schumacher*) *J. Raynal*
63:	I. 1–2	Ascolepis dipsacoides (*Schumacher*) *J. Raynal*
	3–7	Kyllinga erecta *Schumacher*
	II. 1	Kyllinga nemoralis (*Forsk.*) *Dandy ex Hutch.*
	2–7	Kyllinga squamulata *Vahl*
	III. 1–3	Kyllinga squamulata *Vahl*
	4–7	Launaea taraxacifolia (*Willd.*) *Amin ex C. Jeffrey*
64:	I. 1–2	Launaea taraxacifolia (*Willd.*) *Amin ex C. Jeffrey*
	3–7	Lantana camara *L.*
	II. 1	Lantana camara *L.*
	2–3	Leea guineensis *G. Don*
	4	Ophrestia hedysaroides (*Willd.*) *Verdc.*
	5	(Setaria chevalieri *Stapf*
		(Tacca leontopetaloides (*L.*) *Kuntze*

IDC microfiche
No. 2203

64:	II. 6–7	Chlorophytum inornatum *Ker-Gawl.*
	III. 1–4	Eulophia gracilis *Lindl.*
	5–6	Trichilia monadelpha (*Thonning*) *de Wilde*
65:	I. 1–2	Bacopa crenata (*P. Beauv.*) *Hepper*
	3–4	Phragmanthera incana (*Schumacher*) *Balle*
	5	Acridocarpus alternifolius (*Schum. et Thonn.*) *Niedenzu*
	6–7	Manihot esculenta *Crantz*
	II. 1–4	Manihot esculenta *Crantz*
	5–7	Hackelochloa granularis (*L.*) *Kuntze*
	III. 1–2	Hackelochloa granularis (*L.*) *Kuntze*
	3	Marantochloa leucantha (*K. Schum.*) *Milne-Redh.*
	4–7	Marantochloa purpurea (*Ridl.*) *Milne-Redh.*
66:	I. 1–2	Marantochloa purpurea (*Ridl.*) *Milne-Redh.*
	3–7	Mariscus alternifolius *Vahl*
	II. 1–3	Mariscus alternifolius *Vahl*
	4	Marsilea minuta *L.*
	5	Tristemma littorale *Benth.*
	6–7	Melastomastrum capitatum (*Vahl*) *A. et R. Fernandes*
	III. 1	Tristemma hirsutum *P. Beauv.*
	2–4	Dissotis rotundifolia (*Sm.*) *Triana*
	5–7	Tristemma incompletum *R. Br.*
67:	I. 1–2	Tristemma incompletum *R. Br.*
	3–4	Melia azedarach *L.*
	5	Melochia corchorifolia *L.*
	6	Melochia melissifolia *Benth.* var. bracteosa (*F. Hoffm.*) *K. Schum.*
	7	Microlepia speluncae (*L.*) *Moore*
	II. 1–3	Microlepia speluncae (*L.*) *Moore*
	4–7	Albizia adianthifolia (*Schum.*) *W. F. Wight*
	III. 1	Albizia adianthifolia (*Schum.*) *W. F. Wight*
	2–5	Acacia nilotica (*L.*) *Willd.* subsp. adstringens (*Schum. et Thonn.*) *Roberty*
	6–7	Dichrostachys cinerea (*L.*) *Wight et Arn.*
68:	I. 1–2	Dichrostachys cinerea (*L.*) *Wight et Arn.*
	3–6	Mimosa pigra *L.*
	7	Albizia glaberrima (*Schum. et Thonn.*) *Benth.*
	II. 1–4	Albizia glaberrima (*Schum. et Thonn.*) *Benth.*
	5–7	Calliandra portoricensis (*Jacq.*) *Benth.*
	III. 1–2	Calliandra portoricensis (*Jacq.*) *Benth.*
	3–7	Acacia pentagona (*Schum. et Thonn.*) *Hook. f.*
69:	I. 1–4	Mimosa pigra *L.*
	5–7	Acacia pentagona (*Schum. et Thonn.*) *Hook. f.*
	II. 1	Acacia pentagona (*Schum. et Thonn.*) *Hook. f.*
	2–3	Mimosa pudica *L.*
	4	Polycarpaea stellata (*Willd.*) *DC.*
	5–6	Polycarpaea eriantha *Hochst.* ex *A. Rich.* var. effusa (*Oliv.*) *Turrill*
	7	Musanga cecropioides *R. Br.*
	III. 1	Musanga cecropioides *R. Br.*
	2	Cola gigantea *A. Chev.* var. glabrescens *Brenan et Keay*
	3	Mussaenda erythrophylla *Schum. et Thonn.*
	4	Mussaenda isertiana *DC.*
	5–7	Alafia scandens (*Thonning*) *De Wild.*
70:	I. 1–5	Alafia scandens (*Thonning*) *De Wild.*
	6–7	Diospyros tricolor (*Schum. et Thonn.*) *Hiern*

IDC microfiche
No. 2203

70:	II. 1–7	Diospyros tricolor (*Schum. et Thonn.*) *Hiern*
	III. 1	Diospyros tricolor (*Schum. et Thonn.*) *Hiern*
	2–7	Nymphaea guineensis *Schum. et Thonn.*
71:	I. 1–2	Nymphaea guineensis *Schum. et Thonn.*
	3	Ocimum canum *Sims*
	4–7	Borreria filifolia (*Schum. et Thonn.*) *K. Schum.*
	II. 1	Borreria filifolia (*Schum. et Thonn.*) *K. Schum.*
	2–5	Basilicum polystachyon (*L.*) *Moench*
	6–7	Ocimum gratissimum *L.*
	III. 1–7	Ocimum canum *Sims*
72:	I. 1–4	Ocimum canum *Sims*
	5–6	Ocymum barbatum *Vahl*
	7	Orthosiphon suffrutescens (*Thonning*) *J. K. Morton*
	II. 1–3	Orthosiphon suffrutescens (*Thonning*) *J. K. Morton*
	4–7	Endostemon tereticaulis (*Poir.*) *M. Ashby*
	III. 1–2	Endostemon tereticaulis (*Poir.*) *M. Ashby*
	3	Ludwigia abyssinica *A. Rich.*
	4–6	Vahlia dichotoma (*Murr.*) *Kuntze*
	7	Olyra latifolia *L.*
73:	I. 1–2	Olyra latifolia *L.*
	3	Ophioglossum costatum *R. Br.*
	4–7	Habenaria filicornis (*Thonning*) *Lindl.*
	II. 1–4	Habenaria filicornis (*Thonning*) *Lindl.*
	5–7	Urginea ensifolia (*Thonning*) *Hepper*
	III. 1	Urginea ensifolia (*Thonning*) *Hepper*
	2–7	Allophylus spicatus (*Poir.*) *Radlk.*
74:	I. 1–4	Allophylus spicatus (*Poir.*) *Radlk.*
	5–7	Aphania senegalensis (*Juss. ex Poir.*) *Radlk.*
	II. 1–4	Aphania senegalensis (*Juss. ex Poir.*) *Radlk.*
	5–7	Allophylus africanus *P. Beauv.*
	III. 1–7	Allophylus africanus *P. Beauv.*
75:	I. 1	Oxytenanthera abyssinica (*A. Rich.*) *Munro*
	2–3	Pancovia bijuga *Willd.*
	4–7	Pennisetum polystachion (*L.*) *Schult.*
	II. 1	Pennisetum polystachion (*L.*) *Schult.*
	2	Brachiaria falcifera (*Trin.*) *Stapf*
	3–5	Brachiaria deflexa (*Schumacher*) *C. E. Hubbard*
	6	Paspalidium geminatum (*Forsk.*) *Stapf*
	7	Oplismenus hirtellus (*L.*) *P. Beauv.*
	III. 1	Oplismenus hirtellus (*L.*) *P. Beauv.*
	2–3	Brachiaria lata (*Schumacher*) *C. E. Hubbard*
	4	Setaria barbata (*Lam.*) *Kunth*
	5–7	Setaria longiseta *P. Beauv.*
76:	I. 1	Brachiaria mutica (*Forsk.*) *Stapf*
	2–7	Setaria pallide-fusca (*Schumacher*) *Stapf et Hubbard*
	II. 1–4	Setaria pallide-fusca (*Schumacher*) *Stapf et Hubbard*
	5	Panicum brevifolium *L.*
	6–7	Brachiaria distichophylla (*Trin.*) *Stapf*
	III. 1–2	Brachiaria distichophylla (*Trin.*) *Stapf*
	3–6	Panicum maximum *Jacq.*
77:	I. 1–7	Pennisetum subangustum (*Schumacher*) *Stapf et Hubbard*
	II. 1–3	Pennisetum subangustum (*Schumacher*) *Stapf et Hubbard*
	4–7	Setaria sphacelata (*Schumacher*) *Stapf et Hubbard ex M. B. Moss*

IDC microfiche
No. 2203

77: III.	1	Brachiaria distichophylla (*Trin.*) *Stapf*
	2–4	Pavetta corymbosa (*DC.*) *F. N. Williams*
	5	Adenia lobata (*Jacq.*) *Engl.*
	6–7	Paspalum orbiculare *Forst.*
78: I.	1–7	Paullinia pinnata *L.*
II.	1–2	Paullinia pinnata *L.*
	3	Pavetta genipifolia *Schumacher*
	4–5	Pavetta subglabra *Schumacher*
	6	Pennisetum polystachion (*L.*) *Schult.*
	7	Pennisetum purpureum *Schumacher*
III.	1–2	Pennisetum purpureum *Schumacher*
	3–4	Pennisetum americanum (*L.*) *K. Schum.*
	5	Oxystelma bornuense *R. Br.*
	6–7	Asclepiadaceae indet.
79: I.	1–3	Canthium horizontale (*Schumacher*) *Hiern*
	4–6	Glinus oppositifolius (*L.*) *A. DC.*
	7	Polycarpon prostratum (*Forsk.*) *Asch. et Schweinf.*
II.	1	Mollugo cerviana (*L.*) *Seringe*
	2–5	Mollugo nudicaulis *Lam.*
	6	Pharus latifolius *L.*
	7	Phaseolus lunatus *L.*
III.	1	Phaseolus lunatus *L.*
	2–7	Leucas martinicensis (*Jacq.*) *Ait. f.*
80: I.	1–2	Leucas martinicensis (*Jacq.*) *Ait. f.*
	3–4	Leonotis nepetifolia (*L.*) *Ait. f.* var. africana (*P. Beauv.*) *J. K. Morton*
	5–7	Phoenix reclinata *Jacq.*
II.	1–7	Phoenix reclinata *Jacq.*
III.	1	Phoenix reclinata *Jacq.*
	2–6	Phyllanthus amarus *Schum. et Thonn.*
81: I.	1–2	Phyllanthus amarus *Schum. et Thonn.*
	3–7	Securinega virosa (*Roxb. ex Willd.*) *Baill.*
II.	1	Securinega virosa (*Roxb. ex Willd.*) *Baill.*
	2–7	Phyllanthus capillaris *Schum. et Thonn.*
III.	1–5	Phyllanthus capillaris *Schum. et Thonn.*
	6	Securinega virosa (*Roxb. ex Willd.*) *Baill.*
82: I.	1–7	Securinega virosa (*Roxb. ex Willd.*) *Baill.*
II.	1–7	Securinega virosa (*Roxb. ex Willd.*) *Baill.*
III.	1–3	Securinega virosa (*Roxb. ex Willd.*) *Baill.*
	4–6	Phyllanthus pentandrus *Schum. et Thonn.*
83: I.	1–7	Phyllanthus pentandrus *Schum. et Thonn.*
II.	1–5	Phyllanthus pentandrus *Schum. et Thonn.*
	6–7	Phyllanthus reticulatus *Poir.* var. glaber *Müll Arg.*
III.	1–2	Phyllanthus reticulatus *Poir.* var. glaber *Müll Arg.*
	3–4	Phyllanthus amarus *Schum. et Thonn.*
	5–7	Phyllanthus sublanatus *Schum. et Thonn.*
84: I.	1–7	Phyllanthus sublanatus *Schum. et Thonn.*
II.	1–2	Phyllanthus sublanatus *Schum. et Thonn.*
	3–7	Phyllanthus maderaspatensis *L.*
III.	1–2	Phyllanthus maderaspatensis *L.*
	3–4	Microdesmis puberula *Hook. f. ex Planch.*
	5–7	Physalis angulata *L.*
85: I.	1–6	Piper guineense *Thonning*
	7	Pistia stratiotes *L.*

IDC microfiche
No. 2203

85:	II. 1	Pistia stratiotes *L.*
	2–5	Vigna vexillata (*L.*) *A. Rich.*
	6–7	Plumbago zeylanica *L.*
	III. 1–7	Plumbago zeylanica *L.*
86:	I. 1–4	Eragrostis cilianensis (*All.*) *Lut.*
	5–7	Eragrostis ciliaris (*L.*) *R. Br.*
	II. 1	Eragrostis ciliaris (*L.*) *R. Br.*
	2–5	Eragrostis aspera (*Jacq.*) *Nees*
	6–7	Eragrostis linearis (*Schumacher*) *Benth.*
	III. 1–6	Eragrostis linearis (*Schumacher*) *Benth.*
	7	Eragrostis turgida (*Schumacher*) *De Wild.*
87:	I. 1–7	Eragrostis turgida (*Schumacher*) *De Wild.*
	II. 1–3	Eragrostis turgida (*Schumacher*) *De Wild.*
	4–7	Baphia nitida *Lodd.*
	III. 1–2	Baphia nitida *Lodd.*
	3–7	Polygala arenaria *Willd.*
88:	I. 1–7	Polygala arenaria *Willd.*
	II. 1–3	Polygala arenaria *Willd.*
	4–7	Polygala guineensis *Willd.*
	III. 1–5	Polygala guineensis *Willd.*
	6–7	Tectaria angelicifolia (*Schumacher*) *Copel.*
89:	I. 1	Microsorium punctatum (*L.*) *Copel.*
	2	Ctenitis protensa (*Afzel. ex Sw.*) *Ching*
	3–4	Arthropteris palisotii *Desv.*
	5–6	Pouchetia africana *DC.*
	7	Premna quadrifolia *Schum. et Thonn.*
	II. 1–5	Premna quadrifolia *Schum. et Thonn.*
	6	Psidium guajava *L.*
	7	Chassalia kolly (*Schumacher*) *Hepper*
	III. 1	Canthium multiflorum (*Thonning*) *Hiern*
	2–4	Geophila obvallata (*Schumacher*) *F. Didr.*
	5–7	Cremaspora triflora (*Thonning*) *K. Schum.*
90:	I. 1–5	Psychotria calva *Hiern*
	6	Cephaelis peduncularis *Salisb.*
	7	Pteris togoensis *Heiron.*
	II. 1–2	Pteris acanthoneura *Alston*
	3–4	Pteris atrovirens *Willd.*
	5	Pteris acanthoneura *Alston*
	6–7	Pteris atrovirens *Willd.*
	III. 1–2	Pteris atrovirens *Willd.*
	3–5	Pterocarpus santalinoides *L'Hérit. ex DC.*
	6–7	Quisqualis indica *L.*
91:	I. 1–7	Ormocarpum sennoides (*Willd.*) *DC.* subsp. hispidum (*Willd.*) Brenan et *J. Léonard*
	II. 1–7	Ormocarpum sennoides (*Willd.*) *DC.* subsp. hispidum (*Willd.*) Brenan et *J. Léonard*
	III. 1	Rauvolfia vomitoria *Afzel.*
	2–5	Remirea maritima *Aubl.*
	6–7	Dichapetalum guineense (*DC.*) *Keay*
92:	I. 1–4	Dichapetalum guineense (*DC.*) *Keay*
	5–7	Hilleria latifolia (*Lam.*) *H. Walt.*
	II. 1–4	Lonchocarpus sericeus (*Poir.*) *Kunth*
	5–7	Lonchocarpus cyanescens (*Schum. et Thonn.*) *Benth.*

IDC microfiche
No. 2203

92: III.	1–3	Lonchocarpus cyanescens (*Schum. et Thonn.*) *Benth.*
	4–7	Millettia irvinei *Hutch. et Dalz.*
93: I.	1–7	Millettia irvinei *Hutch. et Dalz.*
II.	1	Millettia irvinei *Hutch. et Dalz.*
	2–7	Millettia thonningii (*Schumacher*) *Baker*
III.	1–7	Millettia thonningii (*Schumacher*) *Baker*
94: I.	1–2	Millettia thonningii (*Schumacher*) *Baker*
	3–4	Psychotria calva *Hiern*
	5	Phaulopsis ciliata (*Willd.*) *Hepper*
	6–7	Rhynchospora holoshoenoides (*L. C. Rich.*) *Herter*
II.	1–2	Rhynchospora corymbosa (*L.*) *Britt.*
	3–4	Fimbristylis complanata *Link*
	5–6	Rhynchospora micrantha *Vahl*
	7	Rhynchelytrum repens (*Willd.*) *C. E. Hubbard*
III.	1–3	Rhynchelytrum repens (*Willd.*) *C. E. Hubbard*
	4–6	Sanseveria liberica *Gér. et Labr.*
	7	Scirpus aureiglumis *Hooper*
95: I.	1–2	Scirpus aureiglumis *Hooper*
	3–7	Bulbostylis pilosa (*Willd.*) *Cherm.*
II.	1–7	Bulbostylis pilosa (*Willd.*) *Cherm.*
III.	1	Bulbostylis pilosa (*Willd.*) *Cherm.*
	2–7	Griffonia simplicifolia (*Vahl ex DC.*) *Baill.*
96: I.	1–6	Alchornea cordifolia (*Schum. et Thonn.*) *Müll. Arg.*
	7	Alectra vogelii *Benth.*
II.	1–3	Schwenckia americana *L.*
	4–7	Bulbostylis barbata (*Rottb.*) *C.B.Cl.*
III.	1–4	Bulbostylis barbata (*Rottb.*) *C.B.Cl.*
	5–7	Bulbostylis filamentosa (*Vahl*) *C.B.Cl.*
97: I.	1–5	Bulbostylis filamentosa (*Vahl*) *C.B.Cl.*
	6–7	Fimbristylis hispidula (*Vahl*) *Kunth*
II.	1	Fimbristylis hispidula (*Vahl*) *Kunth*
	2–5	Fimbristylis obtusifolia (*Lam.*) *Kunth*
	6–7	Scleria verrucosa *Willd.*
III.	1–2	Scleria verrucosa *Willd.*
	3–6	Sclerocarpus africanus *Jacq. ex Murr.*
98: I.	1–4	Sclerocarpus africanus *Jacq. ex Murr.*
	5–6	Scoparia dulcis *L.*
	7	Sesamum alatum *Thonning*
II.	1	Sesamum alatum *Thonning*
	2–3	Sesamum radiatum *Schum. et Thonn.*
	4	Sesuvium portulacastrum (*L.*) *L.*
	5	Mitracarpus villosus (*Sw.*) *DC.*
	6–7	Borreria verticillata (*L.*) *G. F. W. Mey.*
III.	1	Setaria anceps *Stapf ex Massey*
	2–5	Pennisetum americanum (*L.*) *K. Schum.*
	6–7	Selaginella myosurus (*Sw.*) *Alston*
99: I.	1–4	Selaginella myosurus (*Sw.*) *Alston*
	5–7	Abutilon guineense (*Schum. et Thonn.*) *Bak. f. et Exell*
II.	1–2	Abutilon guineense (*Schum. et Thonn.*) *Bak. f. et Exell*
	3–7	Sida linifolia *Juss. ex Cav.*
III.	1–3	Sida linifolia *Juss. ex Cav.*
	4–7	Wissadula amplissima (*L.*) *R. E. Fries* var. rostrata (*Schum. et Thonn.*) *R. E. Fries*

IDC microfiche
No. 2203

100: I. 1–6   Wissadula amplissima *R. E. Fries* var. rostrata (*Schum. et Thonn.*)
                R. E. Fries
        7   Sida acuta *Burm. f.*
     II. 1–7   Sida acuta *Burm. f.*
    III. 1–4   Sida alba *L.*
        5–6   Hibiscus micranthus *L. f.*
101: I. 1–2   Sida alba *L.*
        3   Sida cordifolia *L.*
        4–7   Solanum anomalum *Thonning*
     II. 1–7   Solanum anomalum *Thonning*
    III. 1–2   Solanum anomalum *Thonning*
        3–6   Solanum macrocarpon *L.*
102: I. 1–4   Solanum dasyphyllum *Schum. et Thonn.*
        5–7   Solanum indicum *L.* subsp. distichum (*Schum. et Thonn.*) *Bitter*
     II. 1–3   Solanum indicum *L.* subsp. distichum (*Schum. et Thonn.*) *Bitter*
        4–6   Solanum melongena *L.*
        7   Solanum gilo *Raddi* var. gilo
    III. 1–5   Solanum nigrum *L.*
        6–7   Solenostemon monostachyus (*P. Beauv.*) *Briq.* subsp. monostachyus
103: I. 1–7   Solenostemon monostachyus (*P. Beauv.*) *Briq.* subsp. monostachyus
     II. 1–5   Solenostemon monostachyus (*P. Beauv.*) *Briq.* subsp. monostachyus
        6–7   Drepanocarpus lunatus (*L. f.*) *G. F. W. Mey.*
    III. 1–3   Drepanocarpus lunatus (*L. f.*) *G. F. W. Mey.*
        4–5   Sophora occidentalis *L.*
        6–7   Sorghum bicolor (*L.*) *Moench.*
104: I. 1–4   Sorghum bicolor (*L.*) *Moench.*
        5   Sphenoclea zeylanica *Gaertn.*
        6–7   Stachytarpheta angustifolia (*Mill.*) *Vahl*
     II. 1–7   Stachytarpheta angustifolia (*Mill.*) *Vahl*
    III. 1   Stachytarpheta angustifolia (*Mill.*) *Vahl*
        2   Borreria verticillata (*L.*) *G. F. W. Mey.*
        3   Mitracarpus villosus (*Sw.*) *DC.*
        4   Borreria verticillata (*L.*) *G. F. W. Mey.*
        5–7   Mitracarpus villosus (*Sw.*) *DC.*
105: I. 1–2   Borreria sp.
        3–6   Cola verticillata (*Thonning*) *Stapf ex A. Chev.*
        7   Chasmanthera dependens *Hochst.*
     II. 1–3   Mucuna sloanei *Fawc. et Rendle*
        4   Euadenia trifoliolata (*Vahl ex Thonning*) *Oliv.*
        5–7   Ancylobotrys scandens (*Schum. et Thonn.*) *Pichon*
    III. 1–3   Ancylobotrys scandens (*Schum. et Thonn.*) *Pichon*
        4–6   Stylosanthes erecta *P. Beauv.*
106: I. 1–2   Stylosanthes erecta *P. Beauv.*
        3–5   Tacca leontopetaloides (*L.*) *Kuntze*
        6–7   Tephrosia elegans *Schum.*
     II. 1   Tephrosia elegans *Schum.*
        2–4   Tephrosia pumila (*Lam.*) *Pers.*
        5–7   Tephrosia purpurea (*L.*) *Pers.* subsp. leptostachya (*DC.*) *Brummitt*
                    var. leptostachya
    III. 1   Tephrosia purpurea (*L.*) *Pers.* subsp. leptostachya (*DC.*) *Brummitt*
                    var. leptostachya
        2–3   Hippocratea africana (*Willd.*) *Loes. ex Engl.*
        4–6   Thonningia sanguinea *Vahl*

IDC microfiche
No. 2203

107:	I. 1	Trianthema portulacastrum *L.*
	2–3	Tribulus terrestris *L.*
	4–7	Triumfettia rhomboidea *Jacq.*
	II. 1–7	Triumfettia rhomboidea *Jacq.*
	III. 1–3	Triumfettia rhomboidea *Jacq.*
	4–5	Typha domingensis *Pers.*
	6–7	Oxyanthus racemosus (*Schum. et Thonn.*) *Keay*
108:	I. 1–5	Oxyanthus racemosus (*Schum. et Thonn.*) *Keay*
	6	Mitragyna inermis (*Willd.*) *Kuntze*
	7	Conocarpus erectus *L.*
	II. 1–4	Mitragyna inermis (*Willd.*) *Kuntze*
	5–7	Urena lobata *L.*
108:	III. 1–2	Urena lobata *L.*
	3	Usteria guineensis *Willd.*
	4	Utricularia inflexa *Forsk.* var. inflexa
	5–6	Uvaria ovata (*Dunal*) *A. DC.* subsp. ovata
	7	Uvaria chamae *P. Beauv.*
109:	I. 1–3	Uvaria chamae *P. Beauv.*
	4	Annona glauca *Schum. et Thonn.*
	5	Lindernia diffusa (*L.*) *Wettst.* var. diffusa
	6–7	Vigna ambacensis *Bak.*
	II. 1–4	Aspilia ciliata (*Schum.*) *Willd.*
	5	Vetiveria nigritana ((*Benth.*) *Stapf*
	6	Virectaria procumbens (*Sm.*) *Bremek.*
	7	Vitex doniana *Sweet*
	III. 1–5	Vitex doniana *Sweet*
	6–7	Vitex ferruginea *Schum. et Thonn.*
110:	I. 1–4	Vitex ferruginea *Schum. et Thonn.*
	5	Voacanga africana *Stapf*
	6–7	Enydra fluctuans *Lour.*
	II. 1–6	Enydra fluctuans *Lour.*
	7	Waltheria indica *L.*
	III. 1–6	Waltheria indica *L.*
111:	I. 1–4	Waltheria indica *L.*
	5	Berlinia grandiflora (*Vahl*) *Hutch. et Dalz.*
	6–7	Afzelia parviflora (*Vahl*) *Hepper*
	II. 1–2	Wormskioldia pilosa (*Willd.*) *Schweinf. ex Urb.*
	3–4	Ximenia americana *L.*
	5–7	Fagara zanthoxyloides *Lam.*
	III. 1–6	Fagara zanthoxyloides *Lam.*
112:	I. 1–4	Fagara zanthoxyloides *Lam.*
	5–7	Zornia glochidiata *Reichb. ex DC.*
	II. 1–2	Zornia glochidiata *Reichb. ex DC.*

V(d) LIST OF ADDITIONAL ISERT AND THONNING SPECIMENS AT COPENHAGEN
NOT PHOTOGRAPHED FOR IDC MICROFICHE SET 2203.

ACANTHACEAE
Hygrophila abyssinica (*Hochst. ex Nees*) *T. Anders.*

ANACARDIACEAE
Sorindeia warneckei *Engl.*

CAESALPINIACEAE
Piliostigma thonningii (*Schumacher*) *Milne-Redh.* (2 sheets)

COMPOSITAE
Aspilia helianthoides (*Schum. et Thonn.*) *Oliv. et Hiern*

CUCURBITACEAE
Cucumis melo *L.* var. agrestis *Naud.*
Kedrostis foetidissima (*Jacq.*) *Cogn.*
Lagenaria breviflora (*Benth.*) *G. Roberty* (3 sheets)
Luffa aegyptiaca *Mill.*
Momordica foetida *Schumacher*
Zehneria capillacea (*Schum. et Thonn.*) *C. Jeffrey*
Z. hallii *C. Jeffrey*

DILLENIACEAE
Tetracera alnifolia *Willd.*

HYPOXIDACEAE
Curculigo pilosa (*Schum. et Thonn.*) *Engl.*

MALVACEAE
Abelmoschus moschatus *Medik.*
Abutilon guineensis (*Schum. et Thonn.*) *Bak. f. et Exell* (2 sheets)

OCHNACEAE
Ouratea flava (*Schum. et Thonn.*) *Hutch. et Dalz. ex Stapf* (2 sheets)

PAPILIONACEAE
Abrus precatorius *L.* subsp. africanus *Verdc.*
Aeschynomene indica *L.*
Arachis hypogaea *L.*
Stylosanthes erecta *P. Beauv.*
Voandzeia subterranea (*L.*) *Thouars* (3 sheets)

SOLANACEAE
Capsicum frutescens *L.*

STERCULIACEAE
Melochia corchorifolia *L.*

TILIACEAE
Corchorus aestuans *L.*

UMBELLIFERAE
Centella asiatica (*L.*) *Urb.*

VERBENACEAE
Stachytarpheta angustifolia (*Mill.*) *Vahl*

VIOLACEAE
Hybanthus enneaspermus (*L.*) *F.v. Müll.*
H. thesiifolius (*Juss. ex Poir.*) *Hutch.* (3 sheets)

PTERIDOPHYTA
Lygodium microphyllum (*Cav.*) *R. Br.*
Thelypteris elata (*Mett. ex Kuhn*) *Schelpe*

V(e)   INDEX TO MICROFICHE OF ISERT AND THONNING SPECIMENS APPEAR-
       ING ON IDC 2204 "TYPE HERBARIUM OF MUSEUM BOTANICUM
       HAUNIENSE" (COPENHAGEN).

61 : II. 1–4   Cucumis melo *L.* var. agrestis *Naud.*
61 : II. 5–6   Rhaphidiocystis chrysocoma (*Schum.*) *C. Jeffrey*

V(f) LIST OF THONNING SPECIMENS TRACED IN THE WEBB HERBARIUM, UNIVERSITY OF FLORENCE (FI).

NOTE: At the time of listing (March 1972) only a few of the specimens were numbered in preparation for microfilming.

Acacia pentagona (*Schum. et Thonn.*)
   Hook. f.
Allophylus spicatus (*Poir.*) *Radlk.*
Alysicarpus rugosus (*Willd.*) *DC.*
Annona glauca *Schum. et Thonn.*
Aponogeton subconjugatus *Schum. et*
   Thonn.
Borreria filifolia (*Schum. et Thonn.*)
   K. Schum.
Canthium multiflorum (*Thonning*) *Hiern*
Carissa edulis *Vahl*
Cassia absus *L.*
Coldenia procumbens *L.*
Corchorus trilocularis *L.* No. 27790
Crotalaria goreensis *Guill. et Perr.*
Cyperus amabilis *Vahl*
Deinbollia pinnata (*Poir.*) *Schum. et*
   Thonn.
Desmodium ramosissimum *G. Don*
   triflorum (*L.*) *DC.*
   velutinum (*Willd.*) *DC.*
Ehretia corymbosa *Bojer ex A. DC.*
Endostemon tereticaulis (*Poir.*) *M. Ashby*
Erythroxylum emarginatum *Thonning*
   No. 28647
Euphorbia purpurescens *Schum. et Thonn.*
Fimbristylis triflora (*L.*) *K. Schum.*
Gardenia ternifolia *Schum. et Thonn.*
Griffonia simplicifolia (*Vahl ex DC.*)
   Baill.
Gymnema sylvestre (*Retz.*) *Schu tes*
Heliotropium strigosum *Willd.*
Hibiscus micranthus *L. f.* No. 25796
   surattensis *L.* No. 25876
Hippocratea africana (*Willd.*) *Loes.*
   ex Engl.
Hyptis lanceolata *Poir.*
Indigofera dendroides *Jacq.*
   nigricans *Vahl ex Pers.*
   pilosa *Poir.*
   pulchra *Willd.*

secundiflora *Poir.*
spicata *Forsk.*
   subulata *Vahl. ex Poir.* var. subulata
   tetrasperma *Vahl ex Pers.*
Ipomoea coptica (*L.*) *Roth ex Roem. et*
   Schult.
Jacquemontia tamnifolia (*L.*) *Griseb.*
Kohautia virgata (*Willd.*) *Bremek.*
Leucas martinicensis (*Jacq.*) *Ait. f.*
Ludwigia octovalis (*Jacq.*) *Raven* subsp.
   brevisepala (*Brenan*) *Raven*
Macrotyloma biflorum (*Schum. et Thonn.*)
   Hepper
Mariscus ligularis (*L.*) *Urb.*
Merremia tridentata (*L.*) *Hall. f.* subsp.
   angustifolia (*Jacq.*) *Oostr.*
Mitragyna inermis (*Willd.*) *Kuntze*
Ocimum canum *Sims*
Ormocarpum sennoides (*Willd.*) *DC.*
   subsp. hispidum (*Willd.*) *Brenan et*
   J. Léonard
Pergularia daemia (*Forsk.*) *Chiov.*
Phyllanthus amarus *Schum. et Thonn.*
   capillaris *Schum. et Thonn.*
   pentandrus *Schum. et Thonn.*
Polycarpaea stellata (*Willd.*) *DC.*
Pupalia lappacea (*L.*) *Juss.*
Rhynchosia minima (*L.*) *DC.*
Ruspolia hypocrateriformis (*Vahl*)
   Milne-Redh.
Sesbania pachycarpa *DC. emend. Guill. et*
   Perr.
Sida linifolia *Juss. ex Cav.* No. 24930
Solanum anomalum *Thonning*
Uraria picta (*Jacq.*) *DC.*
Utricularia inflexa *Forsk.* var. inflexa
Uvania chamae *P. Beauv.*
   chamae *P. Beauv.*
   ovata (*Vahl ex Dunal*) *A. DC.* subsp.
   ovata
Vitex ferruginea *Schum. et Thonn.*

V(g)   INDEX TO MICROFICHE OF THONNING SPECIMENS TRACED IN THE DE
CANDOLLE HERBARIUM, GENEVA (G-DC).

NOTE: most of the sheets are well written up on their reverse side which is seldom
shown on the microfiche. There must be many others yet to be traced in this
Herbarium.

DC. PRODR. REF. VOL.: PAGE, SP.	CURRENT BOTANICAL NAME (for synonymy see text)	IDC MICROFICHE REF.
1 : 240. 24	Wormskioldia pilosa (*Willd.*) *Schweinf.* ex *Urb.*	121: II.6 (Isert) 121: II.7 (Thonning)
1 : 507. 19	Triumfetta rhomboidea *Jacq.*	nil
2 : 30. 10	Dichapetalum guineense (*DC.*) *Keay*	280: II.3
2 : 86. 7	Connarus thonningii (*DC.*) *Schellenb.*	294: III.7
2 : 132. 95	Crotalaria pallida *Ait.* var. obovata (*G. Don*) *Polhill*	308: II.4
2 : 222. 2	Indigofera tetrasperma *Vahl ex Pers.*	347: II.4
2 : 223. 14	subulata *Vahl ex Poir.*	347: III.7
2 : 227. 59	dendroides *Jacq.*	349: III.4
2 : 228. 65	hirsuta *L.* var. hirsuta	nil
2 : 228. 72	spicata *Forsk.*	350: II.5
2 : 230. 86	pulchra *Willd.*	350: III.3
2 : 500. 128	Cassia absus *L.*	433: I.8
2 : 505. 182β	mimosoides *L.*	434: III.4
3 : 20. 17	Combretum mucronatum *Schumacher*	477: III.4
3 : 20. 21	racemosum *P. Beauv.*	478: I.3
3 : 271. 69	Eugenia coronata *Schum. et Thonn.*	539: I.8
3 : 374. 7	Polycarpaea stellata (*Willd.*) *DC.*	556: II.8
4 : 303. 125	Phragmanthera incana (*Schumacher*) *Balle*	669: II.1
4 : 371. 8	Mussaenda erythrophylla *Schum. et Thonn.*	684: II.6
4 : 372. 19	elegans *Schum. et Thonn.*	685: I.4
4 : 382. 26	Gardenia ternifolia *Schum. et Thonn.*	686: III.2
7 : 138. 37	Launea taraxacifolia (*Willd.*) *Amin* ex *Jeffrey*	1270: III.7
9 : 480. 44	Cordia guineensis *Thonning*	1665: I.8
11 : 633. 13	Premna quadrifolia *Schum. et Thonn.*	1889: II.6
13(1): 259. 617	Solanum anomalum *Thonning*	2081: II.8
13(1): 313. 731	dasyphyllum *Schum. et Thonn.*	2085: III.7
15(2): 337. 177	Phyllanthus pentandrus *Schum. et Thonn.*	2478: I.5
15(2): 338. 179	capillaris *Schum. et Thonn.*	2478: I.8
15(2): 362. 243	maderaspatensis *L.*	2482: I.4

V(h)   THONNING SPECIMENS TRACED IN THE HERBARIUM, ROYAL BOTANIC
GARDENS, KEW (K)

Polycarpaea stellata (*Willd.*) *DC.*
Thonningia sanguinea *Vahl*

V(i) LIST OF THONNING SPECIMENS TRACED IN KOMAROV BOTANICAL
      INSTITUTE HERBARIUM, LENINGRAD (LE).

Many others may be traced here.

Aristida adscensionis *L.* (2 sheets)
  sieberana *Trin.*
Evolvulus alsinoides (*L.*) *L.*
Momordica foetida *Schumacher*
Solanum macrocarpon *L.*
Vitex doniana *Sweet*
Zehneria capillacea (*Schum. et Thonn.*) *C. Jeffrey*
  hallii *C. Jeffrey*

V(j) LIST OF ISERT SPECIMENS TRACED IN BOTANISCHE STAATSSAMLUNG,
      MUNICH (M).

Evolvulus alsinoides (*L.*) *L.*
Rhaphidiocystis chrysocoma (*Schum.*) *C. Jeffrey*
Usteria guineensis *Willd.*

V(k) INDEX TO MICROFICHE OF THONNING SPECIMENS TRACED IN HERBARIUM
      JUSSIEU, JARDIN DES PLANTES, PARIS (P-JU).

NOTE: These sheets are all labelled "e Guinea" or "miss. Vahl 1804", having been
sent to Jussieu by Vahl. Although there is little doubt that they were collected by
Thonning and most of them are types, they are entered in the main part of this work
under the relevant species and cited as "e Guinea" in exsiccatae.

JUSSIEU CATALOGUE NO.	CURRENT BOTANICAL NAME	IDC MICROFICHE NO. 6206
1708	Kyllinga erecta *Schumacher*	112: I.1
1708 bis	squamulata *Vahl*	112: I.1
1721	Bulbostylis pilosa (*Willd.*) *Cherm.*	112: III.2
1779	Fimbristylis hispidula (*Vahl*) *Kunth*	116: I.7
1780	Bulbostylis barbata (*Rottb.*) *C.B.Cl.*	116: II.1
1781	Lipocarpha sphacelata (*Vahl*) *Nees*	116: II.2
1782	Fimbristylis pilosa *Vahl*	116: II.3
1782 bis	Bulbostylis filamentosa (*Vahl*) *C.B.Cl.*	112: I.6
1798	Cyperus margaritaceus *Vahl*	117: III.6, 7
1870 bis	amabilis *Vahl*	122: I.4
1870 ter	dilatatus *Schumacher*	122: I.3
1884 bis	Mariscus ligularis (*L.*) *Urb.*	122: I.2
2016	Fuirena umbellata *Rottb.*	125: III.1
2463	Eragrostis turgida (*Schumacher*) *De Wild.*	156: I.1
2668	Rhynchelytrum repens (*Willd.*) *C. E. Hubbard*	169: III.1
2731	indexed under Andropogon as "Vahl 1804": specimen missing	
2745	Enteropogon macrostachyus (*Hochst. ex A. Rich.*) *Munro ex Benth.*	174: III.6
3209	Commelina erecta *L.*	211: II.5
3244	Iphigenia ledermannii *Engl. et K. Krause*	213: III.1
3479	Haemanthus multiflorus *Martyn*	229: III.3
4390	Hilleria latifolia (*Lam.*) *H. Walt.*	297: III.4

H

JUSSIEU CATALOGUE NO.	CURRENT BOTANICAL NAME	IDC MICROFICHE NO. 6206
4582B	Pupalia lappacea (*L.*) *Juss.*	312: II.3, partly
4590	Pandiaka angustifolia (*Vahl*) *Hepper*	313: I.2
4828	Utricularia inflexa *Forsk.* var. inflexa	331: II.7
4944A	Sesamum radiatum *Schum. et Thonn.*	341: I.5
4954	Spathodea campanulata *P. Beauv.*	342: II.2
5027	Clerodendrum capitatum (*Willd.*) *Schum. et Thonn.* var. capitatum	347: II.5
5052	Vitex ferruginea *Schum. et Thonn.*	349: I.6
5075	Premna quadrifolia *Schum. et Thonn.*	351: I.5
5116	Lantana camara *L.*	354: II.2
5272	Hoslundia opposita *Vahl et Thonn.*	370: II.7
5412	Hyptis lanceolata *Poir.*	382: II.1
5614A	Leucas martinicensis (*Jacq.*) *Ait. f.*	400: 1.4
5694	Endostemon tereticaulis (*Poir.*) *M. Ashby*	406: III.6
5695	Ocimum canum *Sims*	406: III.7
5704	Solenostemon monostachyus (*P. Beauv.*) *Briq.* subsp. monostachyus	407: III.1
5724	Blepharis maderaspatensis (*L.*) *Heyne ex Roth* subsp. rubiifolia *Napper*	408: III.2
5767A	Ruspolia hypocrateriformis (*Vahl*) *Milne-Redh.*	411: III.3
5960	Striga linearifolia (*Schum. et Thonn.*) *Hepper*	425: II.5
6097B	Micrargeria filiformis (*Schum. et Thonn.*) *Hutch. et Dalz.*	436: II.5
6261B	Schwenckia americana *L.*	449: II.7
6376A	Solanum nigrum *L.*	460: III.3
6431	Solanum anomalum *Thonning*	465: III.7
6494	Cordia guineensis *Thonning*	470: III.1
6504	Ehretia corymbosa *Bojer ex A. DC.*	471: II.7
6564	Coldenia procumbens *L.*	476: II.3
6571	Heliotropium strigosum *Willd.*	477: I.6
6738	Jacquemontia ovalifolia (*Vahl*) *Hall. f.*	492: III.2
6760	Ipomoea ochracea (*Lindl.*) *G. Don*	494: I.5
6764	Hewittia sublobata (*L. f.*) *Kuntze*	494: III.1
6766	Ipomoea sepiaria *Roxb.*	494: III.3
6767	Merremia tridentata (*L.*) *Hall. f.*	494: III.4
6803	Jacquemontia tamnifolia (*L.*) *Griseb.*	497: II.7
6845	Ipomoea mauritiana *Jacq.*	500: II.7
6850B	Ipomoea coptica (*L.*) *Roth ex Roem. et Schult.*	501: I.1
6877	Ipomoea stolonifera (*Cyrill.*) *J. F. Gmel.*	503: III.4
6913	Hydrolea glabra *Schum. et Thonn.*	506: II.6
7033	Gymnema sylvestre (*Retz.*) *Schultes*	516: III.6
7057	Leptadenia hastata (*Pers.*) *Decne.*	518: III.3
7061	Pergularia daemia (*Forsk.*) *Chiov.*	518: III.7
7118	Motandra guineensis (*Thonning*) *A. DC.*	523: III.2
7198	Carissa edulis *Vahl*	529: III.4
7235	Synsepalum dulcificum (*Schum. et Thonn.*) *Daniell*	532: I.5
7263	Indexed as "(Chrysophyllum africanum?) Guinée Vahl 1804": specimen missing	
9778	Pavetta corymbosa (*DC.*) *F. N. Williams*	723: I.1
9788	Pavetta genipifolia *Schumacher*	723: II.7
9789	Pavetta subglabra *Schumacher*	723: III.1
9797	Canthium horizontale (*Schumacher*) *Hiern*	724: II.2
9798	Vangueriopsis spinosa (*Schum. et Thonn.*) *Hepper*	724: II.3

JUSSIEU CATALOGUE NO.	CURRENT BOTANICAL NAME	IDC MICROFICHE NO. 6206
9828	Cremaspora triflora (*Thonning*) *K. Schum.*	726: I.6
9829	Chassalia kolly (*Schumacher*) *Hepper*	726: I.7
9830	Canthium multiflorum (*Thonning*) *Hiern*	726: II.1
9851	Kohautia virgata (*Willd.*) *Bremek.*	727: III.2
9853	Mitracarpus villosus (*Sw.*) *DC.*	727: III.4 partly
9854	Diodia sarmentosa *Sw.*	727: III.4 partly
9855	Borreria scabra (*Schum. et Thonn.*) *K. Schum.*	727: III.4
9857	Oldenlandia corymbosa *L.*	727: III.7
9868	Pentodon pentandrus (*Schum. et Thonn.*) *Vatke*	728: II.7
9869	Oldenlandia lancifolia (*Schumacher*) *DC.*	728: II.7
9934	Oxyanthus racemosus (*Schum. et Thonn.*) *Keay*	733: I.3
10015	Mitragyna inermis (*Willd.*) *Kuntze*	738: II.2
10779	Annona senegalensis *Pers.* subsp. senegalensis	799: II.3 partly
10781	glauca *Schum. et Thonn.*	799: II.5
10789	Uvaria chamae *P. Beauv.*	800: I.2
11258	Capparis brassii *DC.*	834: I.1
11272	Indeterminate (see p. 37, under Ritchiea reflexa)	834: III.5
11273	Capparis erythrocarpos *Isert*	834: III.6
11353C	Paullinia pinnata *L.*	841: I.3
11373	Allophylus spicatus (*Poir.*) *Radlk.*	842: I.6
11388	Deinbollia pinnata (*Poir.*) *Schum. et Thonn.*	843: III.1
11455	Erythroxylum emarginatum *Thonning*	849: I.5
11661	Acridocarpus alternifolius (*Schum. et Thonn.*) *Niedenzu*	866: I.3
11675	Triaspis odorata (*Willd.*) *A. Juss.*	867: I.2
11893	Ximenia americana *L.*	882: I.6
11933	Afraegle paniculata (*Schum. et Thonn.*) *Engl.*	884: III.4
11997	Cyphostemma cymosa (*Schum. et Thonn.*) *Descoings*	888: III.3
12032	Hippocratea africana (*Willd.*) *Loes. ex Engl.*	891: I.2
12244	Sida linifolia *Juss. ex Cav.*	907: II.2
12372	Hibiscus surattensis *L.*	916: I.6
12443	Cola verticillata (*Thonning*) *Stapf ex A. Chev.*	921: II.7
12490	Melochia corchorifolia *L.*	924: II.3
12524	Corchorus aestuans *L.*	926: II.5
12528	trilocularis *L.*	926: III.5
12534	tridens *L.*	927: I.5
12542	Triumfetta rhomboidea *Jacq.*	927: III.6
12594	Flacourtia flavescens *Willd.*	931: II.2
12850	Polygala arenaria *Willd.*	949: III.3
12991C	Fagara zanthoxyloides *Lam.*	959: I.2
13018(5)	Ouratea flava (*Schum. et Thonn.*) *Hutch. et Dalz. ex Stapf*	961: III.4
13367	Polycarpaea stellata (*Willd.*) *DC.*	989: III.4
13400	Gisekia pharnaceoides *L.*	992: II.2
13640	Quisqualis indica *L.*	1008: III.2
13745	Ludwigia octovalvis (*Jacq.*) *Raven* subsp. brevisepala (*Brenan*) *Raven*	1015: I.4
13939	Eugenia coronata *Schum. et Thonn.*	1026: III.6
14051	Dissotis rotundifolia (*Sm.*) *Triana*	1033: III.5
14083	Tristemma incompletum *R. Br.*	1035: III.1
14438	Albizia glaberrima (*Schum. et Thonn.*) *Benth.*	1061: III.1
14469	Calliandra portoricensis (*Jacq.*) *Benth.*	1064: III.4

JUSSIEU CATALOGUE NO.	CURRENT BOTANICAL NAME	IDC MICROFICHE NO. 6206
14472	Albizia adianthifolia (*Schumacher*) *W. F. Wight*	1064: I.7
14478	Acacia nilotica (*L.*) *Willd. ex Del.* subsp. adstringens (*Schum. et Thonn.*) *Roberty*	1064: II.7
14505	Acacia pentagona (*Schum. et Thonn.*) *Hook. f.*	1066: I.7
14532	Cassia absus *L.*	1068: I.7
14622	Griffonia simplicifolia (*Vahl ex DC.*) *Baill.*	1074: III.1
14701	Baphia nitida *Lodd.*	1080: II.1
14907	Crotalaria glauca *Willd.*	1093: III.3?
14924B	goreensis *Guill. et Perr.*	1094: III.3
14925	pallida *Ait.* var. obovata (*G. Don*) *Polhill*	1094: III.4
15150	Canavalia rosea (*Sw.*) *DC.*	1113: II.1
15157	virosa (*Roxb.*) *Wight et Arn.*	1113: III.1
15177	Eriosema glomeratum (*Guill. et Perr.*) *Hook. f.*	1116: 1.6
15179	Macrotyloma biflorum (*Schum. et Thonn.*) *Hepper*	1116: II.3
15180	Rhynchosia sublobata (*Schum. et Thonn.*) *Meikle*	1116: II.4
15215	minima (*L.*) *DC.*	1119: II.1
15224	Lonchocarpus cyanescens (*Schum. et Thonn.*) *Benth.*	1120: I.6
15225	sericeus (*Poir.*) *Kunth*	1120: I.7
15226	Millettia thonningii (*Schumacher*) *Baker*	1120: II.1
15356	Tephrosia pumila (*Lam.*) *Pers.*	1129: I.1
15357	purpurea (*L.*) *Pers.* subsp. leptostachya (*DC.*) *Brummitt* var. leptostachya	1129: I.2
15365A	Indigofera tetrasperma *Vahl ex Pers.*	1129: II.7
15366	paniculata *Vahl ex Pers.*	1129: III.2
15371	nigricans *Vahl ex Pers.*	1130: I.3
15376	spicata *Forsk.*	1130: II.4
15378	pulchra *Willd.*	1130: II.6
15382	dendroides *Jacq.*	1130: III.3
15383	secundiflora *Poir.*	1130: III.4
15384	subulata *Vahl ex Poir.* var. subulata	1130: III.5
15385	macrophylla *Schumacher*	1130: III.6
15386	hirsuta *L.* var. hirsuta	1130: III.7
15498	Alycarpus ovalifolius (*Schum. et Thonn.*) *J. Léonard*	1139: III.4
15512	Sesbania pachycarpa *DC. emend. Guill. et Perr.*	1141: I.2
15513	sericea (*Willd.*) *Link*	1141: I.3
15575	Alysicarpus rugosus (*Willd.*) *DC.*	1145: III.6
15576	Desmodium triflorum (*L.*) *DC.*	1145: III.7
15577	gangeticum (*L.*) *DC.*	1146: I.1
15578	velutinum (*Willd.*) *DC.*	1146: I.2
15579	Uvaria picta (*Jacq.*) *DC.*	1146 I.3
15580	Desmodium ramosissimum *G. Don*	1146: I.4
15636	Stylosanthes erecta *P. Beauv.*	1149: III.3
15973	Clausena anisata (*Willd.*) *Hook. f. ex Benth.*	1159: III.7
15977A	Cnestis ferruginea *DC.*	1160: 1.4
16238	Phyllanthus maderaspatensis *L.*	1176: III.3
16239	pentandrus *Schum. et Thonn.*	1176: III.4
16303	capillaris *Schum. et Thonn.*	1180: III.4
16305	sublanatus *Schum. et Thonn.*	1180: III.6
16333	Securinega virosa (*Roxb. ex Willd.*) *Baill.*	1182: III.2
16390	Euphorbia purpurascens *Schum. et Thonn.*	1185: III.4
16392	prostrata *Ait.*	1185: III.6

JUSSIEU		IDC MICROFICHE
CATALOGUE NO.	CURRENT BOTANICAL NAME	NO. 6206
16524A	Alchornea cordifolia (*Schum. et Thonn.*)	
	Müll. Arg.	1196: I.5
16537	Acalypha ciliata *Forsk.*	1197: I.1
16585	Mallotus oppositifolius (*Geisel.*) *Müll. Arg.*	1200: I.5
16606	Zehneria hallii *C. Jeffrey*	1201: II.7
16622	Cucumis melo *L.* var. agrestis *Naud.*	1202: III.5
17374	Aponogeton subconjugatus *Schum. et Thonn.*	1255: II.4

V(l) LIST OF ISERT AND THONNING SPECIMENS TRACED IN NATURHISTORIKA
RIKSMUSEUM, STOCKHOLM (S).

Abutilon guineense (*Schum. et Thonn.*) *Bak. f. et Exell*
Adenia lobata (*Jacq.*) *Engl.* (2 sheets)
Aspilia hilianthoides (*Schum. et Thonn.*) *Oliv. et Hiern*
Brachiaria deflexa (*Schumacher*) *C. E. Hubbard*
    distichophylla (*Trin.*) *Stapf*
    lata (*Schumacher*) *C. E. Hubbard*
Cremaspora triflora (*Thonning*) *K. Schum.*
Crotalaria glauca *Willd.*
    pallida *Ait.* var. obovata (*G. Don*) *Polhill*
Curculigo pilosa (*Schum. et Thonn.*) *Engl.*
Eclipta alba (*L.*) *Hassk.*
Endostemon tereticaulis (*Poir.*) *M. Ashby*
Evolvulus alsinoides (*L.*) *L.* (2 sheets)
Ficus exasperata *Vahl* (2 sheets)
    ovata *Vahl*
Gossypium herbaceum *L.* var. acerifolium (*Guill. et Perr.*) *A. Chev.*
Gymnema sylvestre (*Retz.*) *Schultes*
Hibiscus cannabinus *L.*
    micranthus *L. f.*
    owariensis *P. Beauv.*
    vitifolius *L.*
Hoslundia opposita *Vahl et Thonn.*
Indigofera hirsuta *L.* var. hirsuta
    macrophylla *Schum.*
    pilosa *Poir.*
    secundiflora *Poir.*
    subulata *Vahl ex Poir.* var. subulata
    tetrasperma *Vahl ex Pers.*
Kohautia virgata (*Willd.*) *Bremek.*
Launaea taraxacifolia (*Willd.*) *Amin ex C. Jeffrey*
Leptadenia hastata (*Pers.*) *Decne.* (3 sheets)
Mitragyna inermis (*Willd.*) *Kuntze*
Mussaenda elegans *Schum. et Thonn.*
Ocimum canum *Sims*
Oldenlandia corymbosa *L.*
    lancifolia (*Schumacher*) *DC.*
Pandiaka angustifolia (*Vahl*) *Hepper*
Panicum brevifolium *L.*
    maximum *Jacq.*
Pavetta subglabra *Schumacher* (2 sheets)
Pentodon pentandrus (*Schum. et Thonn.*) *Vatke*

Piliostigma thonningii (*Schumacher*) *Milne-Redh.*
Polycarpaea eriantha *Hochst. ex A. Rich.* var. effusa (*Oliv.*) *Turrill*
Rhynchelytrum repens (*Willd.*) *C. E. Hubbard*
Setaria longiseta *P. Beauv.*
Sida acuta *Burm. f.*
  alba *L.*
Spilanthes filicaulis (*Schum. et Thonn.*) *C. D. Adams*
Wissadula amplissima *R. E. Fries* var. rostrata (*Schum. et Thonn.*) *R. E. Fries*

INDEX OF RECOGNISED SPECIES
AND SYNONYMS

Plant names mentioned in Schumacher's taxonomic notes are not indexed.